Futuristic Research in Modeling of Dynamical Systems (FRMDS 2022)

Journal of Physics: Conference Series Volume 2332

Online
3-5 May 2022

ISBN: 978-1-7138-6176-8
ISSN: 1742-6588

Printed from e-media with permission by:

Curran Associates, Inc.
57 Morehouse Lane
Red Hook, NY 12571

Some format issues inherent in the e-media version may also appear in this print version.

Institute of Physics
Dirac House, Temple Back
Bristol BS1 6BE UK

Phone: 44 1 17 929 7481
Fax: 44 1 17 920 0979

techtracking@iop.org

Additional copies of this publication are available from:

Curran Associates, Inc.
57 Morehouse Lane
Red Hook, NY 12571 USA
Phone: 845-758-0400
Fax: 845-758-2633
Email: curran@proceedings.com
Web: www.proceedings.com

TABLE OF CONTENTS

PREFACE

International E-conference on Futuristic Research in Modeling of Dynamical Systems (FRMDS 2022) was held on 4th and 5th May 2022, organized by the Department of Mathematics, CMR Institute of Technology, Bengaluru-560037, India. FRMDS 2022 provided a forum for academicians, students, and research scholars to present their innovative ideas, models, applications, research accomplishments, and concepts of diversified areas. The theme of the conference was an emerging interdisciplinary area-dynamical system in various fields of research. Hence, a wide range of theories, methodologies & algorithms were explored among the participants during the conference.

FRMDS 2022 was held online with the help of the Zoom platform, to avoid the spread of corona virus infection. The Conveners were: **Dr. Nidhi S Vaishnaw and Prof. ShankarNarayan S**. During the conference, the keynote speech was presented by an expert member from Academics **Dr. S Sivasankaran**, Department of Mathematics, King Abdulaziz University, KSA. This conference was modeled with two tracks namely: Mathematical modeling and optimization techniques as track1, and Data Science and Machine learning-related topics as track 2. The two-day conference comprised 6 plenarysessions, with 1 speaker session. Speaker was given 50 minutes to present, followed by 5-10 minutes for questions. The conference was structured to foster discussion between participants around the core themes of the Dynamical Systems. This was achieved by hosting small group debates, minute papers, and full group discussions on pre-prepared questions following each session. Each day of the conference also allotted one hour for lunch, and 30 minutes for morning and afternoon tea to allow participants to continue their discussions after each session. Time was allocated at the end of each day for a groupreflection on the day's discussions. More than 35 research works were presented by various participants.

ORGANIZING INSTITUTE LOGO

FRMDS COMMITTEES

A. ORGANIZING TEAM

1. Prof. (Dr.) K Meenakshi (Organizing Chair), CMRIT, Bengaluru
2. Prof. (Dr.) Nidhi S Vaishnaw (Convener), CMRIT, Bengaluru
3. Prof. Shankar Narayan S (Convener), CMRIT, Bengaluru

B. ORGANIZING COMMITTEE

1. Prof. (Dr.) Hanumantha Ravi, Department of Mathematics, CMRIT, Bengaluru
2. Prof. (Dr.) Rajendra P, Department of Mathematics, CMRIT, Bengaluru
3. Prof. (Dr.) Shweta Gaur, Department of Mathematics, CMRIT, Bengaluru
4. Prof. (Dr.) Prathap D, Department of Mathematics, CMRIT, Bengaluru
5. Prof. (Dr.) Shiva Kumar, Department of Mathematics, CMRIT, Bengaluru
6. Prof. Uma Raju, Department of Mathematics, CMRIT, Bengaluru
7. Prof. Asha K N, Department of Mathematics, CMRIT, Bengaluru
8. Prof. Bharti Sharma, Department of Mathematics, CMRIT, Bengaluru

C. ADVISORY COMMITTEE MEMBERS

1. Dr. Arun Kumar T, Department of Mechanical Engineering, CMRIT, Bengaluru
2. Dr. M. Sarada, Department of Mathematics, REVA University, Bengaluru
3. Dr. Pankaj Srivatsava, Department of Mathematics, M.N. National Institute of Technology, Prayagraj
4. Dr. Edwin Raj, Department of Mathematics, Hindustan University, Chennai
5. Dr. Ramdas Balan, Department of Physics, CMRIT, Bengaluru

Declaration:

Name: Dr. Nidhi S Vaishnaw	**Name:** S. Shankar Narayan
Affiliation: CMR Institute of Technology, Bengaluru	**Affiliation:** CMR Institute of Technology, Bengaluru
Email Address: nidhi.v@cmrit.ac.in	**Email Address:** shankar.n@cmrit.ac.in

Peer Review Statement

All papers published in this volume have been reviewed through processes administered by the Editors. Reviews were conducted by expert referees to the professional and scientific standards expected of a proceedings journal published by IOP Publishing Publishing.

1. Type of peer review: Double anonymous
2. Conference submission management system: Morressier
3. Number of submissions received: 39
4. Number of submissions sent for review: 39
5. Number of submissions accepted: 20
6. Acceptance Rate (Submissions Accepted / Submissions Received × 100): 51.28 %
7. Average number of reviews per paper: 2.00
8. Total number of reviewers involved: 24
9. Contact person for queries:
   Name: Dr. Nidhi S Vaishnaw
   Affiliation: CMR Institute of technology
   Email: nidhi.v@cmrit.ac.in

Exponential sum of the first kind

E Kiran Babu[1] , V Raghavendra Prasad[1] and Y Rajasekhara Goud[2]

[1] Assistant Professor of Mathematics, Department of H & BS, G Pulla Reddy Engineering College (Autonomous), Kurnool, INDIA.
[2] Professor of Mathematics, Department of H & BS, G Pulla Reddy Engineering College (Autonomous), Kurnool , INDIA.

Abstract: This paper is mainly concerned with a new exponential sum which unifies Unitary Ramanujan sum and Unitary Nagell's Totient function. A formula for this sum and Holder equality are obtained using the structure of certain classes of integers.

Key words: Ramanujan sum, Mobious inverse formula and Unitary Divisor.

1. INTRODUCTION

In the theory of arithmetic functions the Unitary Divisor concept plays an important role .For instance J. Hanumanthachari [1] has introduced the concept of Boolean divisor motivated by the Unitary Divisor concept for the construction of a non-associative ring of arithmetic functions.

E. Cohen [2] has studied the trigonometric sum (exponential sum) ${C^*}_N(L)$ is defined by the sum of nth power all primitive N roots of Unity. It can be expressed explicitly as in the following definition.

The Unitary Ramanujan sum (URS) defined by

$$C^*(L,N) = \sum_{\substack{1\le x\le N \\ (x,N)^*=1}} exp\left({}^{2\pi i L x}/_{N}\right)$$

where $N, L \in N$ and with fixed N, the Unitary Ramanujan sum ${C^*}_N(L)$ is periodic variable L more over N is a period of ${C^*}_N(L)$. Where $(x,N)^*$ denotes the largest divisor of x which is Unitary divisor of N. It is proved that $C^*(L,N)$ is always an integer and is given by

$$C^*(L,N) = \sum_{d\,||\,(L,N)^*} d\mu^*\left({}^{N}/_{d}\right)$$

may be called as a "divisor sum of Unitary Ramanujan sum."

Also $C^*(L,N) = \varphi^*(g)\mu^*\left({}^{N}/_{g}\right) = \varphi^*(g)C^*(L/g, N/g)$, where g = $(L,N)^*$

The above equation may be called as Holder identity. From the above it can be observed that Unitary Ramanujan sum as exponential sum and divisor sum assume that same values and also connects $\varphi^*(N)$,the Unitary analogue of the Euler Totient function.

2. STRUCTURE OF CLASS OF INTEGERS

THEOREM 2.1: (Euler Structure Theorem)

If d//N and $(b,d)^* = 1$, then the numbers $a \equiv b(mod\ d)$ but in congruent modulo N and $(a,N)^* = 1$ are given by $a_k + t_k\frac{N}{d}$, where a_k runs through UNRRS (mod N/d) and 't_k' is the Solution of congruence $\lambda_k\frac{N}{d} \equiv b - a_k(mod\ d)$.

Proof: Let $(a,N)^* = 1$ and a$\equiv b(Mod\ d)$
Then $(a,d)^* = 1$ and $(a,d)^* = (b,d)^*$;
Further $\left(a,\frac{N}{d}\right)^* = 1$.
Let a$\equiv a_k\left(mod\frac{N}{d}\right)$ then $\left(a_k,\frac{N}{d}\right)^* = 1$, write a = $a_k + \lambda_k\frac{N}{d}$.
Substituting this in the equation a$\equiv b(Mod\ d)$, then $a_k + \lambda_k\frac{N}{d} \equiv b(mod\ d)$
That is $\lambda_k\frac{N}{d} \equiv (b - a_k)(mod\ d)$.
Since (d, N/d) =1, this has unique solution modulo d (say $t_k(mod\ d)$).
Conversely, let a = $a_k + t_k\frac{N}{d}$, where $\left(a_k,\frac{N}{d}\right)^* = 1, 1 \le a_k \le \frac{N}{d}$ and t_k is given above theorem.
Then clearly $(a,N)^* = 1$ and a$\equiv b(Mod\ d)$.
Finally $a_r + t_r\frac{N}{d}, a_s + t_s\frac{N}{d}$ satisfy the following conditions,

$$1 \le a_r, a_s \le \frac{N}{d}, 1 \le t_r, t_s \le d, \left(a_r,\frac{N}{d}\right)^* = 1 = \left(a_s,\frac{N}{d}\right)^*,$$

$$a_r + t_r\frac{N}{d} \equiv b(mod\ d), a_s + t_s\frac{N}{d} \equiv b(mod\ d),$$

$$\left(a_r + t_r\frac{N}{d}, N\right)^* = 1 = \left(a_s + t_s\frac{N}{d}, N\right)^*$$

Suppose $a_r + t_r\frac{N}{d} \equiv a_s + t_s\frac{N}{d}(mod\ N)$ then

$a_r - a_s \equiv t_r - t_s\frac{N}{d}(mod\ N) \equiv 0(mod\ N/d)$

which leads to contradiction unless $a_r - a_s = 0$ or $a_r = a_s$.

Similarly, $t_r = t_s$.

Example (2.2):

Choose, N = 20,d = 5 and b =4.
Note that in the data $(4,5)^* = 1$
Now $(a_k,4)^* = 1 \Rightarrow a_k = 1,2 and\ 3$.
The corresponding $t_k = 2,3,4$.
Therefore $a_k + t_k\frac{N}{d} = 1 + 2.4, 2 + 3.4\ and\ 3 + 4.4$
= 9,14 and 19
Now observe that $9,14,19 \equiv 4(mod\ 5)$ and
$(9,20)^* = (14,20)^* = (19,20)^* = 1$

3. UNITARY RAMANUJAN SUMS OF FIRST KIND

The following lemmas are required to find A formula for C(L,M,N).

LEMMA 3.1: (INCLUSION - EXCLUSION PRINCIPLE)

Let be a finite set of given objects of which the objects having a given property 'a' from the subset A_a those having property 'b' from subset A_b ,though having both a and b from subset A_{ab}, ... those having a, b and c from the subset A_{abc}, ... and so on.

Let f(x) be any real for complex valued function defined on A and let
$S = \sum_{x\in A} f(x)$, $S_a = \sum_{x\in A_a} f(x)$, $S_{ab} = \sum_{x\in A_{ab}} f(x)$ and so on .

LEMMA 3.2: (A MOBIUS INVERSION FORMULA)

If f (x) is a periodic for real x with period 1 , then

$$\sum_{h\in URRS\,(mod\ N)} f(h/N) = \sum_{d||N} \mu^*(N/d) \sum_{a\in CRS(mod\ d)} f(a/d)$$

Proof: Consider the set of N rational fractions $\frac{h}{N}(1 \le h \le N)$

We can express each of these fractions in unitarily irreducible form in a way that $\frac{h}{N} = \frac{a}{d}$, where d||N and $(1 \le a \le d)$, $(a,d)^* = 1$ and 'a' & 'd' are Uniquely determined by 'h' and 'N'.

Conversely, Let $\frac{h}{N} = \frac{a}{d}$, where d||N and $(1 \le a \le d)$, where $(a,d)^* = 1$ and 'a' & 'd' are Uniquely determined by 'h' and 'N'.

Then for any function *f(x)* we have,

$$\sum_{(1\le h\le N)} f(h/N) = \sum_{d||N} \sum_{\substack{(1\le a\le d)\\(a,d)^*=1}} f(a/d)$$

Write $G(N) = \sum_{(1\le h\le N)} f(h/N)$, $F(N) = \sum_{\substack{(1\le h\le N)\\(h,N)^*}} f(h/N)$

Then it becomes $G(N) = \sum_{d||N} F(d)$ which implies that

$$F(N) = \sum_{d||N} \mu^*(N/d)\, G(d)$$

That is

$$\sum_{\substack{(1\le h\le N)\\(h,N)^*}} f(h/N) = \sum_{d||N} \mu^*(N/d) \sum_{(1\le a\le d)} f(a/d)$$

THEOREM 3.3: A **Formula for C(L,M,N):**

The Unitary Nagell –Ramanujan sum C(L,M,N) is given by

$$C(L,M,N) = \sum_{\substack{d||N\\(M,d)^*=1}} \mu^*(d) \sum_{\delta||\frac{N}{d}} \mu^*(N/d\delta)\, J_{L<M<N}(d,\delta)$$

where

$$J_{L<M<N}(d,\delta) = \sum_{a\in CRS} exp\left\{2\pi i L\left(\frac{a}{\delta} + Y_{\frac{a}{\delta}}\right)\frac{N}{d}/N\right\}$$

$Y_{\frac{a}{\delta}}$ Satisfies the congruence $Y_{\frac{a}{\delta}}\frac{N}{d} \equiv M - \frac{a}{\delta}\frac{N}{d}(mod\ d)$.

Proof : By the definition

UNRRS (Mod N, M) = $\{x(mod\ N): (X,N)^* = 1 = (M-X,N)^*\}$

If we remove those $x(mod\ N)$ for which $(M-X,N)^* \ne 1$ from URRS (mod N) ,we get UNRRS (Mod N, M).

Let $N = p_1^{\alpha_1}, p_2^{\alpha_2}, \ldots \ldots . p_t^{\alpha_t}$ and $\{x(mod\ N): (X,N)^* = 1\}$

$A_{p_i} = \{x(mod\ N): (X,N)^* = 1, M - X \equiv 0(mod\ p_i^{\alpha_i})\}$, i = 1,2,3.....t.

$A_{p_ip_j} = \{x(mod\ N): (X,N)^* = 1, M - X \equiv 0(mod\ p_i^{\alpha_i} p_j^{\alpha_j})\}$, i≠ j; j = 1,2,....t. and so on.

Using lemma 1,it follows that

$$C(L,M,N) = \sum_{x\in A} exp(2\pi iLx/N) - \sum_{x\in A_{p_1}} exp(2\pi iLx/N) \dots\dots - \sum_{x\in A_{p_t}} exp(2\pi iLx/N)$$

$$+ \sum_{x\in A_{p_1p_2}} exp(2\pi iLx/N) \dots\dots + (-1)^t \sum_{x\in A_{p_1p_2\dots\dots}} exp(2\pi iLx/N)$$

$$= \sum_{d||N} \mu^*(d) \sum_{\substack{x\in URRS(mod\ N)\\ x\equiv M\equiv(mod\ d)}} exp(2\pi iLx/N)$$

In the above sum $(M,d)^* = 1$.

Hence $C(L,M,N) = \sum_{\substack{d||N\\ (M,d)^*=1}} \mu^*(d) \sum_{\substack{x\in URRS(mod\ N)\\ x\equiv M\equiv(mod\ d)}} exp(2\pi iLx/N)$

Now $(M,d)^* = 1$,

$$\sum_{\substack{x\in URRS(mod\ N)\\ x\equiv M\equiv(mod\ d)}} exp(2\pi iLx/N) = \sum_{\substack{a_k\in URRS\left(mod\frac{N}{d}\right)\\ t_k\frac{N}{d}\equiv M-a_k(mod\ d)}} exp\left\{2\pi iL\left(a_k + t_k\frac{N}{d}\right)/N\right\}$$

For any real number x, define f(x) as follows

$$f(x) = \begin{cases} exp\left\{\frac{2\pi iL}{N}(x+y_x)\frac{N}{d}\right\}, & if\ \frac{xN}{d}\ is\ an\ integer \\ o, & otherwise \end{cases}$$

where y_x is the solution in the variable 'y' of the congruence

$$(x+y)\frac{N}{d} \equiv M(mod\ d).$$

Then 'f' is well defined on the real number system and is periodic with period 1.

On using the definition of 'f', it follows that

$$\sum_{\substack{a_k\in URRS\left(mod\frac{N}{d}\right)\\ t_k\frac{N}{d}\equiv M-a_k(mod\ d)}} exp\left\{2\pi iL\left(a_k + t_k\frac{N}{d}\right)/N\right\} = \sum_{a\in URRS\left(mod\frac{N}{d}\right)} f\left(a_k/\frac{N}{d}\right)$$

$$\sum_{\delta||\frac{N}{d}} \mu^*(N/d\delta) \sum_{a\in CRS(mod\ \delta)} f(a/\delta)$$

Denote $J_{L,M,N} = \sum_{a\in CRS(mod\ \delta)} f(a/\delta)$

$$= \sum_{a\in CRS(mod\ \delta)} exp\left\{\frac{2\pi iL}{N}(a/\delta + y_{a/\delta})\frac{N}{d}\right\}$$

where $y_{a/\delta}\frac{N}{d} \equiv M - a/\delta\frac{N}{d}(mod\ d)$.

Example (3.4):

Choose, L=3, M= 5, N=4

Then $C(3,5,4) = \sum_{x\in UNRRS\ (mod\ 4,5)} exp\left(\frac{2\pi i3x}{4}\right)$

Now , UNRRS (mod 4,5) = (2,3)

There fore $C(3,5,4) = -1 + i$

The divisors of 4 which may satisfy the conditions Viz d// 4 and $(5,d)^* = 1$ are 1 and 4 .

$$C(3,5,4) = \sum_{\substack{d||N \\ (M,d)^*=1}} \mu^*(d) \sum_{\delta||\frac{N}{d}} \mu^*\left(\frac{N}{\delta d}\right) J(d,\delta)$$

= -J (1,1)+J(1,4) –J(4,1)
= -1+0+i
= -1+i

4. References:

1. J Hanumanthachari: Certain generalizations of Nagell's Totient function and Ramanujan's sum, Math. Student 38(1970) 183-187: correction and addition, ibid. 42 (1974)121-124.

2. E Cohen: Arithmetical functions associated with the unitary divisors of an integer, Math. Z.74.(1960) 66-80.

3. G.H.Hardy: Note on Ramanujan's trigonometrical function $C_q(n)$ and certain series of arithmetical functions, Proc. comb. philos. soc. 20(1921) 263-271.

4. P.Vijayasekhara Reddy: Thesis of "Structures of certain class of integers and applications to some Totient functions and allied Ramanujan sums".

Fuzzy inventory model without shortages using GMI approach

P Vasanthi[1], S Ranganayaki[2] and R Kasthuri[3]

[1,2,3] Department of Mathematics, Sri Ramakrishna Engineering College, Coimbatore-641022,TamilNadu,India.

vasanthi.maths@srec.ac.in

Abstract. The aim of this research work is to come out with an inventory model without shortages involving set-up cost and holding cost as imprecise variables. The uncertainty is chosen as fuzzy number in triangularized form and the graded mean integration method is applied to defuzzify the parameters. The effect of defuzzification is explained by analyzing for different input values through numerical examples.

Key words. Inventory model without shortages, Setup cost, Holding cost, Fuzzy variables, GMI method.

1. Introduction

Many researches have been done with inventory model under fuzzy sense in which the various costs and input parameters are fuzzified using different fuzzy number like triangular, trapezoidal and pentagon fuzzy numbers. Bellman and Zadeh [1] first introduced the concept of fuzziness in decision making problems. Yao and Chang [2] developed an EOQ model with fuzzy back order quantity. Hariga [10] determined an optimal EOQ model for deteriorating items with time varying demand whereas Heng et al [11] introduced an order level lot-size inventory model for deteriorating items. Park [3] considered a fuzzy set theoretic approach of economic order quantity and Yao & Lee [4] studied a fuzzy inventory model without backorder for fuzzy order quantity.

Vijusevic considered an inventory model with ordering cost as fuzzy triangular number and holding cost as fuzzy triangular number and determined the annual cost function using centroid defuzzification approach.

In this paper the order cost and holding cost are assumed as triangularized fuzzy number and the graded mean integration process is used to defuzzify imprecise variables.

2. An Inventory model without shortages

In this chapter, an inventory model without shortages has been formulated with order and holding cost under fuzzy sense. The fuzzy parameters are taken as triangularized fuzzy number and an imprecise total cost value is estimated by defuzzifying it by GMI method.

3. Assumptions & Notations

3.1. Assumptions
(i) No shortages
(ii) Replenishment is Instantaneous
(iii) Zero lead –time
(iv) Constant Demand

3.2. Notations
T-length of the plan
c-order cost of an item
s-unit storage cost
v-demand in (0,t)
tQ- length of a cycle
Q-order quantity level
$\tilde{c}$ -fuzzy order cost of an item
$\tilde{s}$-fuzzy unit storage cost
F-Total cost

4. Mathematical Model in crisp sense

As an initiative measure the economic order quantity formula considered by Harris is assumed here which is stated as follows

$$Q^* = \sqrt{\frac{2cv}{sT}} \tag{1}$$

which minimizes the total annual cost value.
Total annual inventory cost = Storage cost + setup cost

$$F = \frac{sTQ}{2} + \frac{cv}{Q} \tag{2}$$

5. Mathematical model under fuzzy sense

The fuzziness in the inventory model is explained by fuzzy set theory in many articles. It is found that few crisp parameters are converted to fuzzy numbers to make the model more realistic.
The mathematical model in fuzzy sense is arrived with modification of the crisp model by introducing the uncertain parametric values. In general, the set-up cost and storage cost are not certain and hence inturn they are taken as fuzzy parameters.

Equation (2) can be rewritten as $\tilde{F} = \frac{\tilde{s}TQ}{2} + \frac{\tilde{c}v}{Q}$ (3)

In this the □ represents the fuzzy parameters. GMI method is used here to determine the optimal solution.
Due to many factors it is difficult to describe the order cost for a long period.
And hence it is valid to consider the ordering cost in an interval of the form
$(c-\alpha_1, c+\alpha_2)$where $o<\alpha_1<c$ and $\alpha_1, \alpha_2>0$.
Suitably the triangular fuzzy number for c is given by
$\tilde{c}=(c-\alpha_1,c,c+\alpha_2)$
The GMI formula for $\tilde{c}$ is

$$\theta(\tilde{c})-=\frac{1}{6}[c-\alpha_1+4c+c+\alpha_2]$$

$$=\text{c}+\frac{1}{6}[\alpha_2-\alpha_1]$$

In real life it is known that the storage price may vary from its actual one in time t. If it lies in the interval ($s-\alpha_3$, $s+\alpha_4$), then similar to ordering cost we can define the storage cost also as a triangular fuzzy number to explain its vagueness.

$\tilde{s}$=($s-\alpha_3$,s,$s+\alpha_4$)

$0<\alpha_3<s,\alpha_3,\alpha_4>0$

The graded mean representation of $\tilde{s}$ is

$$\theta(\tilde{s})=\frac{1}{6}[\text{s}-\alpha_3+4s+s+\alpha_4]$$

$$=s+\frac{1}{6}[\alpha_4-\alpha_3]$$

From equation (3), we have

$$F_Q(c,s)=\frac{sTQ}{2}+\frac{cv}{Q} \tag{4}$$

The fuzzified form of total cost function can be indicated by $F_Q(c.s)=\left[\frac{TQ}{2}\otimes\tilde{s}\right]\oplus\left[\frac{v}{Q}\otimes\tilde{c}\right]$

(5)

where Q>0,T>0 and v>0. By the concept of addition and multiplication of triangular fuzzy number equation (5) can be written as

$$F_Q(\tilde{c},\tilde{s})=\left[F_1,F_2,F_3\right] \tag{6}$$

Here $F_Q(\tilde{c},\tilde{s})$ is the expected fuzzy total cost function

The terms F_1,F_2,F_3 can be estimated as

$$F_1=\left\{F_Q(c,s)-\left[\frac{TQ\alpha_3}{2}+\frac{v\alpha_1}{Q}\right]\right\}$$

$$F_2=F_Q(c,s)=\left[\frac{TQs}{2}+\frac{cv}{Q}\right]$$

$$F_3=F_Q(c,s)+\left[\frac{TQ\alpha_4}{2}+\frac{v\alpha_2}{Q}\right]$$

Defuzzifying the triangularised fuzzy numbers $F_Q(\tilde{c},\tilde{s})$ by the method of graded mean integration results in the following equations

$$\theta\left[F_Q(\tilde{c},\tilde{s})\right]=\frac{1}{6}\left\{F_Q-\left(\frac{TQ\alpha_3}{2}+\frac{v\alpha_1}{Q}\right)\right\}+\ 4F_Q+F_Q+\frac{1}{6}\left\{\frac{TQ\alpha_4}{2}+\frac{v\alpha_2}{Q}\right\}$$

$$\theta\left[F_Q(\tilde{c},\tilde{s})\right]=F_Q+\frac{TQ}{12}\left[\alpha_4-\alpha_3\right]\ +\frac{v}{6Q}[\alpha_2-\alpha_1]$$

$$\theta\left[F_Q\right]=\frac{TQs}{2}+\frac{vc}{Q}+\frac{TQ}{12}\left[\alpha_4-\alpha_3\right]\ +\frac{v}{6Q}[\alpha_2-\alpha_1]$$

$$\frac{d}{dQ}\left[F_Q\right]=0$$

provides the fuzzy value of order quantity as

$$Q^*=\sqrt{\frac{2v(6c+\alpha_2-\alpha_1)}{T(6s+\alpha_4-\alpha_3)}}$$

Also it is clear that

$$\frac{d^2}{dQ^2}(F_Q)=\frac{2vc}{Q^3}+\frac{v}{3Q^3}(\alpha_2-\alpha_1)>0$$

and hence $F_Q(c,s)$ is minimum by the concept of maxima and minima.

6. Numerical solution and its Analysis

6.1. *Crisp values of the model*

Let order cost c=150 per unit, holding cost s= 0.25 per unit, Demand v=60,000 and length of the plan T=6. Then the economic order quantity is Q*=3464 and the total cost value F=5196.15.

6.2. *Fuzzy values of the model*

Input values

Table 1. Triangular fuzzy numbers of order cost (c) and holding cost (s)

$\tilde{C}$	$\tilde{S}$
(45,150,180)	(0.11,0.25,0.38)
(60,150,225)	(0.09,0.25,0.39)
(75,150,210)	(0.1,0.25,0.39)

Output Values

Table 2. Calculated values of order quantity Q* and total cost function F*

Q^*	F^*
3327.76	4954.99
3458.25	5121.43
3347.02	5284.96

The determined minimum fuzzy cost by GMI technique is comparatively less than the crisp values. The estimated economic order quantity by GMI method is nearer to the crisp value.

7. Conclusion

The conventional study of this paper concentrates on optimizing an inventory model in which the set up and holding cost are described as fuzzy numbers. For each of the fuzzy input parameters, the associated optimal order quantity and total annual cost are calculated. By applying the GMI method, a least total cost is obtained. Though the optimal value is unique in crisp model, the fuzzy model provides a minimum solution that optimizes the cost function which will be more helpful in real life situations. The difference between the calculated values of crisp and fuzzy model seems minimum which could be reasonable if we consider higher values for the order and the holding cost.

8. References

[1] Bellman R E and Zadeh L A 1970 Decision making in a fuzzy environment *Management Science*, 17(4),141–164.

[2] Chiang J, Yao J S and Lee H M 2005 Fuzzy inventory with backorder defuzzification by signed distance method *Journal of Information Science and Engineering* Vol **21** pp 673–694.

[3] Park K S 1987 Fuzzy set theoretic interpretation of economic order quantity', *IIIE* Transactions on systems man and cybernetics **17** 1082–84.

[4] H.M.Lee and J.S.Yao 1999 Economic Order Quantity in Fuzzy Sense for Inventory without backorder *Fuzzy sets and Systems* **105** pp 13–31.

[5] Vujosevic M, Petrovic D and Petrovic R 1996 EOQ formula when inventory cost is fuzzy *International Journal Production Economics* **45** pp 499–50.

[6] Ranganayaki S, Kasthuri R andVasanthi P 2019 Inventory model with demanddependent on Unit price under fuzzy parameters and Decision variables *International Journal of Recent Technology and Engineering* **8** pp 784–788.

[7] Syed J K and Aziz L 2007 A Fuzzy Inventory model without shortages using signed distance method *Applied Mathematics and Information Sciences* **1** 203-209.

[8] Dutta D and Pavankumar 2012 Fuzzy Inventory Model without shortages using trapezoidal fuzzy numbers with sensitivity analysis IOSR-Journal of mathematics **4** 32–37.

[9] Kasthuri R Vasanthi P and Ranganayaki S 2020 Optimization of Inventory Model-Cost Parameters Inventory and Lot-size as Fuzzy Numbers *International Journal of Engneering and Advanced Technology* **10** pp 365–368.

[10] Moncer Hariga 1996 Optimal EOQ Models for deteriorating items with time-varying demand *Journal of Operational Research Socie*ty **47** pp 1228-46.

[11] Heng K J Jihad Labban and Richard J L 1991 An Order-level Lot-Size Inventory model for deteriorating items with finite replenishment rate *Computers and Industrial Engineering* **20** pp 187-197.

A fuzzy purchase model with backorders - GMI method of defuzzification

R Kasthuri[1] P Vasanthi[2] S Ranganayaki[3] and K Kavithamani[4]

[1,2,3] Department of Mathematics, Sri Ramakrishna Engineering College, Coimbatore-641022,TamilNadu,India.

[4] Department of Mathematics, Hindusthan College of Engineering and Technology, Coimbatore, TamilNadu, India.

kasthuri.maths@srec.ac.in

Abstract. A fuzzy EOQ model with backorders is considered in which the costs like setup, holding and the penalty price are assumed as triangular FN (fuzzy numbers). Graded mean integration approach which is more simple and accurate is employed to defuzzify the cost functions. The model is illustrated with numerical examples to compute the optimal values.

Key words. Fuzzy Model, GMI, Back orders, Fuzzy number

1. Introduction

Many research articles over inventory management and decision making have been published which presents a number of models describing various conditions and assumptions. Inspite of this, the concept of fuzziness plays a significant part in Operations research. Jain [2] developed inventory model with fuzzified input parameters. Park [3] interpreted an EOQ model with fuzzy set theoretic approach.

Lee and Yao [5] discussed an inventory model with an uncertain demand and production. Karwowski and Evans [6] applied fuzzy concepts in production management while Zimmermann [7] developed fuzzy mathematical programming.

Syed and Aziz [9] dealt with a inventory model without backorders using signed distance method in which triangular fuzzy numbers are made use of the represent the cost parameters in fuzzy sense.

This paper reveals a Purchase inventory model with fuzzified holding cost, setup cost and penalty cost as triangular FN. A method of differential calculus is applied to attain optimum solution of the objective function namely the total cost function. A mathematical model for crisp and fuzzy case are described with a numerical solution. GMI representation of defuzzification has been employed to attain the solution.

2. Fuzzy Set Theory

The impreciseness that governs the decision making process are represented using a powerful tool called fuzzy set theory. The input variables are encountered as triangular functions or trapezoidal functions to describe the uncertainty of the parameters. Due to fuzzy environment the optimal value of the objective function constitute an alternative better solution.

3. Method of Graded Mean Integration

For decision makers, it is more comfortable to use crisp values rather than fuzzy and hence is modeler attempt to convert a fuzzy value to a crisp one. In this work, the process of defuzzification of fuzzy cost function is done using Graded mean integration technique.

Let $\tilde{A} = (b_1, b_2, b_3)$ indicates triangular FN. The GMI representation of fuzzy numbers are estimated as

$$\theta(\tilde{A}) = \frac{1}{6}[b_1 + 4b_2 + b_3]$$

4. Mathematical Model representation

4.1. Crisp Model

The cost function is

$$TC = \frac{SD}{Q} + \frac{I^2H}{2Q} + \frac{(Q-I)^2m}{2Q} \quad (1)$$

$$TC = \frac{SD}{Q} + \frac{I^2(H+m)}{2Q} - Im + \frac{mQ}{2} \quad (2)$$

where Q - lot size, I - allowable inventory, S - order cost, D – uniform demand, H - holding price per and m - backorder cost per item. The economical value of Q is estimated as

$$Q = \sqrt{\frac{2SD+I^2(H+m)}{m}} \quad (3)$$

4.2.*Fuzzy model*

The objective function namely the total cost function with the input parameters described in this chapter are fuzzified entirely.In this chapter, the crisp model presented by equation (1) is fully fuzzified by fuzzifying the input parameters S, H and m and also the objective function namely the total cost function.

The input cost parameters are represented by triangular FN as follows.

$$\tilde{S} = (S - \Delta_1, S, S + \Delta_2)$$
$$\tilde{H} = (H - \Delta_3, H, H + \Delta_4)$$
$$\tilde{m} = (m - \Delta_5, m, m + \Delta_6)$$

where $\Delta_i > 0, i = 1,2,3,4,5,6.$

Decision makers can determine these values of Δ_i as per the inventory system.

$$\widetilde{TC} = \frac{\tilde{S}D}{Q} + \frac{I^2\tilde{H}}{2Q} + \frac{(Q-I)^2\tilde{m}}{2Q} = F_Q \quad (4)$$

That is $\widetilde{TC} = (F_1, F_2, F_3)$

where $F_1 = \frac{(S-\Delta_1)D}{Q} + \frac{I^2(H-\Delta_3)}{2Q} + \frac{(Q-I)^2(m-\Delta_5)}{2Q}$

$$= F_Q - \left\{\frac{\Delta_1 D}{Q} - \frac{I^2\Delta_3}{2Q} - \frac{(Q-I)^2\Delta_5}{2Q}\right\}$$

$F_Q = \frac{SD}{Q} + \frac{I^2H}{2Q} + \frac{(Q-I)^2 m}{2Q}$ = FQ

$$F_3 = \frac{(S+\Delta_2)D}{Q} + \frac{I^2(H+\Delta_4)}{2Q} + \frac{(Q-I)^2(m+\Delta_6)}{2Q}$$
$$= F_Q + \left\{\frac{\Delta_2 D}{Q} + \frac{I^2\Delta_4}{2Q} + \frac{(Q-I)^2\Delta_6}{2Q}\right\}$$

where F_1, F_2 and F_3 are the components of triangular fuzzy numbers.

By GMI representation of $\widetilde{TC}$, we have

$$\theta(\widetilde{TC}) = \frac{1}{6}[F_1 + 4F_2 + F_3]$$
$$= \frac{1}{6}\left[F_Q - \left\{\frac{\Delta_1 D}{Q} - \frac{I^2\Delta_3}{2Q} - \frac{(Q-I)^2\Delta_5}{2Q}\right\}\right] + \frac{2}{3}F_Q + \frac{1}{6}\left[F_Q + \left\{\frac{\Delta_2 D}{Q} + \frac{I^2\Delta_4}{2Q} + \frac{(Q-I)^2\Delta_6}{2Q}\right\}\right]$$
$$= F_Q + \frac{1}{6}\left\{\frac{D(\Delta_2-\Delta_1)}{Q} + \frac{I^2(\Delta_4-\Delta_3)}{2Q} + \frac{(Q-I)^2(\Delta_6-\Delta_5)}{2Q}\right\} \quad (5)$$

Hence

$$\theta(\widetilde{TC}) = \frac{SD}{Q} + \frac{I^2(H+m)}{2Q} - Im + \frac{mQ}{2} + \frac{D(\Delta_2-\Delta_1)}{6Q} + \frac{I^2(\Delta_4-\Delta_3)}{12Q} + \frac{(Q-I)^2(\Delta_6-\Delta_5)}{12Q} \quad (6)$$

The differential calculus concept is used to optimize the fuzzy objective function whose condition is $\frac{\partial \widetilde{TC}}{\partial Q} = 0$ which gives the lot size value as

$$Q = \sqrt{\frac{12SD + 6I^2(H+m) + 2D(\Delta_2-\Delta_1) + I^2(\Delta_4-\Delta_3+\Delta_6-\Delta_5)}{6m+\Delta_6-\Delta_5}} \quad (7)$$

5. Numerical Analysis

The behavior of the developed model has been illustrated here. The computed crisp and fuzzy values are correlated for better results.

Input values

$\tilde{S} = (30.0,150.0,195.0)$ $\widetilde{H} = (0.06,0.25,0.36)$ $\widetilde{m} = (1.0,5.0,7.6)$

and D=60,000

Crisp values

S=137.5 H=0.237 m=4.767

The results obtained are Q*=8887.21 and the corresponding annual cost function is TC=1931.63

Fuzzy values

Q*=8886.95 and $\widetilde{TC} = 1930.28$

Depending upon any number of input fuzzified values and the corresponding crisp values the objective function values can be calculated and from the above numerical analysis it can be concluded that the fuzzy value is better than the crisp value.

6. Conclusion

Present research work deals with a purchase model with decision variables as fuzzy parameters namely triangular FN. The model is solved for both crisp and fuzzy values in order to calculate the annual inventory cost function. The results obtained from the numerical example shows that the estimated fuzzy value is more precise when compared to the crisp value. The difference between the fuzzy and crisp values will be significant if we move with the higher order of investment. It can be concluded that the model can be developed for different types of fuzzy numbers and the values can be compared to attain a better result.

7. References

[1] Zadeh L A and Bellman R E 1970 Decision making in fuzzy environment*Management Science* 17pp 140–164.

[2] Jain R 1976 Decision making in the presence of fuzzy variables*IIIETransactions on Systems, Man and Cybermatics***6**pp 698–703.

[3] Park 1987 Fuzzy set theoretic interpretation of economic order quantity*IIIE Transactions on Systems, Man and Cybermatics* **7** pp 1082–84.

[4] Yao J S and Chiang J 2003 Inventory without backorderwith fuzzy total cost and fuzzy storing cost defuzzified by centroid and signed distance*EuropeanJournal of Operational Research***148** pp 401–409.

[5] LeeH M and Yao J S 1998 Economic production quantity for fuzzy demand quantity and fuzzy production quantity*European Journal of Operations Research***109** pp 203–211.

[6] Karwowski Y and Evans G W 1986 Fuzzy concepts in productionmanagementresearch:a review *International Journal of Production Research***24**pp129–147.

[7] Zimmermann H J 1983 Fuzzy Mathematical Programming *Computers andOperations Research***10** pp 291–298.

[8] Vasanthi P, Ranganayaki S and Kasthuri R 2019 Fuzzy EOQ model withshortages using Kuhn-Tucker conditions *International Journal ofEngineeringand Advanced Technology* **8**pp 822–827.

[9] Sywd J K and Aziz A 2007 Fuzzy Inventory model without shortages usingsigned distance method*Applied Mathematics and Information Sciences***1**pp 203–209.

[10] Yao J S and Lee H M 1999 Fuzzy Inventory model with or without backorderfor fuzzy order quantity with trapezoidal fuzzy number *Fuzzy setsand Systems***105**pp 311–337.

[11] Kazemi N , Ehsani E and Jaber M Y 2010 An Inventory with backorders with fuzzy parameters and decision variables*", International Journal of Approximate Reason* **51** pp 964–972.

[12] Parcathi P and Gajalakshmi S 2013 An Inventory model with allowable shortage using trapezoidal fuzzy numbers*International Journal of Scientific Engineering Research* **4**pp 1068-1073.

ON FNS - COMPACTNESS IN FUZZY NEUTROSOPHIC SUPRA TOPOLOGICAL SPACES

V Amarendra Babu[1,4] and J Aswini[2,5], K Moses[3,6]

[1] Department of Mathematics, Acharya Nagarjuna University, Nagarjuna Nagar, A.P- India,
[2] Department of Mathematics, Acharya Nagarjuna University, Nagarjuna Nagar, A.P- India,
[3] Department of Mathematics, Andhra Christian Degree & PG Collage Guntur, A.P- India,
[4]amarendrab4@gmail.com
[5]aswinijampana9@gmail.com
[6]kommalapati26@gmail.com

Abstract: In this paper, we introducing fuzzy neutrosophic supra first countable (FNS - FCS), fuzzy neutrosophic supra second countable (FNS- SCS) and fuzzy neutrosophic supra compactness (FNS- compactness), fuzzy neutrosophic supra Q^{μ} - compactness (FNS - Q^{μ} - compactness) in fuzzy neutrosophic supra topological space (FNSTS). we derive the union of two FNS – compact spaces is also FNS – compact space and similarly the union of two FNS -Q^{μ} - compact spaces is also FNS - Q^{μ} - compact space. Also we define some theorems using finite intersection property and productive property. Finally we observe that our notions preserve under one – one, onto and continuous mapping.

1. Introduction: In 1965, Zadeh [9] introduced the notion of fuzzy sets. The study of fuzzy topological spaces was first initiated by chang [4] in 1968. K. Atanassov [8] introduced the notion of intuitionistic fuzzy sets (Ifs) in 1986. Later Coker [5,6] introduced intuitionistic fuzzy topological space. In 1983[2], Mashhour et al. introduced the concepts of supra topological spaces, supra open sets& supra closed sets. Later on 1987, ME Abd El-Monsef et al [10] introduced the concept of fuzzy supra topological as a natural generalization of the notion of supra topological spaces.

In 2002, Florentin Smarandache [15] introduced the extend the IFsets into Nsets. Nset is classified into 3 independent functions. In 2012, Salama, Alblowi [1,13,14] introduced the NTY. NTSs are very natural generalizations of FTSs. In 2014, Salama, Smarandache and valeri [14] introduced the concept of neutrosophic closed sets and continuous functions. Dogen Coker [3,7]& Bayhan introduced Fuzzy-compactness in IFTSs. In 2018,2019,[11,12] Md.Aman Mahbub introduced some properties of compactness in IFTSs and On Q-Compactness in IFTSs.

In this paper two definitions of FNS - compactness in 'fuzzy 'neutrosophic supra 'topological 'space and some of their features are defined.

2. Preliminaries:

2.1. Definition:[15]Let X be a nonempty set and P is a fuzzy neutrosophic set (FNS) is an object having the form $P=\{\langle l,T_P(l),I_P(l),F_P(l)\rangle : l\in X\}$,where the functions

$T_P : X \to]^-0,1^+[$, $I_P : X \to]^-0,1^+[$, $F_P : X \to]^-0,1^+[$

denote the degree of membership function $(T_P(l))$, the degree of indeterminacy function ($I_P(l)$), and the degree of non membership ($F_P(l)$) respectively, each element $l\in X$ to the set P and $^+0\le T_P(l)\le I_P(l)\le F_P(l)\le 1^+$, for each $l\in X$.

A FNS $P=\{\langle l,T_P(l),I_P(l),F_P(l)\rangle : l\in X\}$ can be identified to an ordered triple $\langle l,T_P,I_P,F_P\rangle$ on $]^-0,1^+[$ on X.

2.2. Definition: [1,15] Let X be a nonempty set and the 'FNSs P & R be in the form $P=\{\langle l,T_P(l),I_P(l),F_P(l)\rangle : l\in X\}$ & $R=\{\langle l,T_R(l),I_R(l),F_R(l)\rangle : l\in X\}$ in X

1. $CO(P)=\{\langle 1-T_P(l),1-I_P(l),1-F_P(l)\rangle : l\in X\}$
2. $P\subseteq R \Leftrightarrow T_P(l)\le T_R(l), I_P(l)\le I_R(l), F_P(l)\ge F_R(l)$, for all $l\in X$,
3. $P-R=\{\langle l,T_P(l)\wedge F_R(l),I_P(l)\wedge(1-I_R(l)),F_P(l)\vee T_R(l)\rangle : l\in X\}$
4. $[]P=\{\langle l,T_P(l),I_P(l),1-T_P(l)\rangle : l\in X\}$
5. $\langle\ \rangle P=\{\langle l,1-F_P(l),I_P(l),F_P(l)\rangle : l\in X\}$
6. $0_{FN}=\langle l,0,0,1\rangle$ and $1_{FN}=\langle l,1,1,0\rangle$

2.3. Definition: [15] Let $\{P_j : j\in K\}$be an arbitrary family of FNSs in X, where $P_j=\{\langle l,T_{P_j}(l),I_{P_j}(l),F_{P_j}(l)\rangle : l\in X\}$ then

1. $\cap P_j=\{\langle l,\wedge_{j\in K}T_{P_j}(l),\wedge_{j\in K}I_{P_j}(l),\vee_{j\in K}F_{P_j}(l)\rangle : l\in X\}$
2. $\cup P_j=\{\langle l,\vee_{j\in K}T_{P_j}(l),\vee_{j\in K}I_{P_j}(l),\wedge_{j\in K}F_{P_j}(l)\rangle : l\in X\}$

2.4. Definition: [13,17] A fuzzy neutrosophic supra topology (FNST), Let X be a nonempty set & τ^{μ} be a family of FNS subsets in X, satisfying the following axioms.

$0_{FN}, 1_{FN} \in \tau^{\mu}$

$\cup M_i \in \tau^{\mu}, \forall \{M_i : i \in K\} \subseteq \tau^{\mu}$

In the pair, (X, τ^{μ}) is said to be a Fuzzy neutrosophic supra topological space (FNSTS) and any FNSS in τ^{μ} is known as Fuzzy neutrosophic supra open set (FNSOS) in X. The element of τ^{μ} are called FNSOSs. The complement of FNSOS in the FNSTS (X, τ^{μ}) is called fuzzy neutrosophic supra closed set (FNSCS).

2.5. Definition: [3] Let M and N are IFSs on X and Y. Then the product of IFSs M and N denoted by $M \times N$ is defined by $M \times N = \{(r,t), T_M \times T_N, F_M \times F_N\}$ where $(T_M \times T_N)(r,t) = \min(T_M(r), T_N(t))$ and $(F_M \times F_N)(r,t) = \max(F_M(r), F_N(t))$ for all $(r,t) \in X \times Y$. Obviously $0 \le (T_M \times T_N) + (F_M \times F_N) \le 1$. This definition can be extended to an arbitrary family of IFSs.

2.6 .Definition: [3] Let $(X_i, Y_i), i = 1,2$ be two IFTSs. The product topology $\tau_1 \times \tau_2$ on $X_1 \times X_2$ is the IFT generated by $\{p_i^{-1}(V_i) : V_i \in \tau_i, i = 1,2\}$ where $p_i : X_1 \times X_2 \to X_i, i = 1,2$ are the projection maps and IFTS $\{X_1 \times X_2, \tau_1 \times \tau_2\}$ is called the product IFTS of $(X_i, Y_i), i = 1,2$. In this case $\delta = \{p_i^{-1}(V_i) : V_i \in \tau_i, i \in K\}$ is a sub base and $B = \{V_1 \times V_2 : V_i \in \tau_i, i = 1,2\}$ is base for $\tau_1 \times \tau_2$ in $X_1 \times X_2$

2.7. Definition: [14] (i) If $M = \langle T_M, I_M, F_M \rangle$ is a NS on Y, then the preimage of M under *h*, denoted by $h^{-1}(M)$, is a NS in X defined by $h^{-1}(M) = \langle h^{-1}(T_M), h^{-1}(I_M).h^{-1}(F_M) \rangle$.
(ii) If $N = \langle T_N, I_N, F_N \rangle$ is a NS in X, then the image of N under *h*, denoted by $h(N)$, is a NS in Y defined by $h(N) = \langle h(T_N), h(I_N).h(F_N) \rangle$.

2.8 .Definition: [14] Let $(X, \tau^{\mu}), (Y, \sigma^{\mu})$ be FNTSs. A function $h : X \to Y$ is called neutrosophic Continuous, if $h^{-1}(M) \in \tau^{\mu}$ for all $M \in \sigma^{\mu}$ and *h* is called Neutrosophic Open, if $h(N) \in \sigma^{\mu}$ for all $N \in \tau^{\mu}$.

2.9. Definition: [16] A collection ϑ of NS on a set X is called basis (or base) for a NTS on X, if
(i) For each $p_N^x \in X$, there exists $E \in \vartheta$ such that $p_N^x \in E$,
(ii) If $p_N^x \in E_1 \cap E_2$, where $E_1, E_2 \in \vartheta$ then $\exists E_3 \in \vartheta$ such that $p_N^x \in E_3 \subseteq E_1 \cap E_2$.

2.10. Definition: [11,12](i) Let (X, τ^{μ}) be a NTS and E be a NS on X. If a family $\{\langle l, T_{E_i}, I_{E_i}, F_{E_i} \rangle : l \in K\}$ of NOSs on X satisfies the condition, $\cup \{\langle l, T_{E_i}, I_{E_i}, F_{E_i} \rangle : l \in K\} = 1_N$, then it is said to be a neutrosophic open cover of X. A finite sub family of neutrosophic open cover

$\{\langle l, T_{E_l}, I_{E_l}, F_{E_l}\rangle : l \in K\}$ of X, which is also NOC of X is called a finite sub cover of $\{\langle l, T_{E_l}, I_{E_l}, F_{E_l}\rangle : l \in K\}$.

3. FNS – First and Second countable space in FNSTS

3.1. Definition: A FNSTS (X, τ^{μ}) is called fuzzy neutrosophic supra first countable space (FNS – FCS) if for every FNP P_{FN} there exists a countable local base.

3.2. Example: Let $X = \{l, m, n\}$ and $\tau^{\mu} = \{0_{FN}, 1_{FN}, M_1, M_2, M_3\}$, where

$$M_1 = \left\{\left\langle \frac{l}{0.25}, \frac{l}{0.45}, \frac{l}{0.75}\right\rangle\left\langle \frac{m}{0.28}, \frac{m}{0.45}, \frac{m}{0.72}\right\rangle\left\langle \frac{n}{0.35}, \frac{n}{0.55}, \frac{n}{0.65}\right\rangle\right\},$$

$$M_2 = \left\{\left\langle \frac{l}{0.45}, \frac{l}{0.45}, \frac{l}{0.55}\right\rangle\left\langle \frac{m}{0.48}, \frac{m}{0.45}, \frac{m}{0.52}\right\rangle\left\langle \frac{n}{0.55}, \frac{n}{0.55}, \frac{n}{0.45}\right\rangle\right\}$$ and

$$M_3 = \left\{\left\langle \frac{l}{0.55}, \frac{l}{0.45}, \frac{l}{0.45}\right\rangle\left\langle \frac{m}{0.65}, \frac{m}{0.45}, \frac{m}{0.35}\right\rangle\left\langle \frac{n}{0.75}, \frac{n}{0.55}, \frac{n}{0.25}\right\rangle\right\}.$$ Then (X, τ^{μ}) is a FNSTS.

Let $$P_{FN} = \left\{\left\langle \frac{l}{0.35}, \frac{l}{0.45}, \frac{l}{0.65}\right\rangle\left\langle \frac{m}{0.38}, \frac{m}{0.45}, \frac{m}{0.62}\right\rangle\left\langle \frac{n}{0.45}, \frac{n}{0.55}, \frac{n}{0.55}\right\rangle\right\}$$ be a FNP.

$\mathrm{H}(P_{FN}) = \{V_1, V_2\}$, where $$V_1 = \left\{\left\langle \frac{l}{0.45}, \frac{l}{0.45}, \frac{l}{0.55}\right\rangle\left\langle \frac{m}{0.48}, \frac{m}{0.45}, \frac{m}{0.52}\right\rangle\left\langle \frac{n}{0.55}, \frac{n}{0.55}, \frac{n}{0.45}\right\rangle\right\},$$

$$V_2 = \left\{\left\langle \frac{l}{0.55}, \frac{l}{0.45}, \frac{l}{0.45}\right\rangle\left\langle \frac{m}{0.65}, \frac{m}{0.45}, \frac{m}{0.35}\right\rangle\left\langle \frac{n}{0.75}, \frac{n}{0.55}, \frac{n}{0.25}\right\rangle\right\}.$$

Write $$\mathrm{N}^* = \left\{\left\langle \frac{l}{0.85}, \frac{l}{0.55}, \frac{l}{0.25}\right\rangle\left\langle \frac{m}{0.88}, \frac{m}{0.55}, \frac{m}{0.22}\right\rangle\left\langle \frac{n}{0.95}, \frac{n}{0.55}, \frac{n}{0.15}\right\rangle\right\}$$ is a FNSNHD of FNP P_{FN}.

Then $V_1, V_2 \subseteq \mathrm{N}^*$. Therefore $\mathrm{H}(P_{FN})$ is local base of FNP P_{FN}. Hence (X, τ^{μ}) is FNS – FCS.

3.3. Theorem: Let (X, τ^{μ}) & (Y, δ^{μ}) are any FNSTSs and $h : X \to Y$ is onto Open and FNS – continuous mapping. If (X, τ^{μ}) is FNS – FCS then (Y, δ^{μ}) is also FNS – FCS.

Proof: Suppose $P_{FN}^{\ y}$ is a FNP on Y. Since $h : (X, \tau^{\mu}) \to (Y, \delta^{\mu})$ is onto then $\exists$ FNP $P_{FN}^{\ y} \in Y \ni h(P_{FN}^{\ x}) = P_{FN}^{\ y}$. Since (X, τ^{μ}) is FNS – FCS then $\exists$ a countable local base. Say H of $P_{FN}^{\ x}$. We have to prove that *h* (H) has a countable local base in Y at $P_{FN}^{\ y}$. Since H is countable, so *h* (H) is countable. Let N be a FNSNHD of $P_{FN}^{\ y}$.

Since *h* is FNS – continuous then h^{-1} (N) $\in \tau^{\mu}$ such that $P_{FN}^{\ x} \in h^{-1}$ (N) & $\exists$ $\mathrm{H}^* \in$ H such that $P_{FN}^{\ x} \in \mathrm{H}^* \subseteq$

h^{-1} (N). Since h is open, $h(P_{FN}{}^{x}) \in h$ (H) $\subseteq$ N. So $P_{FN}{}^{y} \in h(\mathrm{H}^{*}) \subseteq$ G. Hence h (H) is countable local base for Y. Therefore (Y, δ^{μ}) is also FNS – FCS.

3.4. Definition: A FNSTS (X, τ^{μ}) is called fuzzy neutrosophic supra second countable space (FNS – SCS) if it has a countable base.

3.5. Example: Let $X = \{p, q, r\}$ and $\tau^{\mu} = \{0_{FN}, K, L, M, 1_{FN}\}$, where

$$K = \left\{\left\langle \frac{l}{0.24}, \frac{l}{0.45}, \frac{l}{0.76} \right\rangle \left\langle \frac{m}{0.34}, \frac{m}{0.45}, \frac{m}{0.66} \right\rangle \left\langle \frac{n}{0.44}, \frac{n}{0.55}, \frac{n}{0.56} \right\rangle\right\},$$

$$L = \left\{\left\langle \frac{l}{0.46}, \frac{l}{0.45}, \frac{l}{0.54} \right\rangle \left\langle \frac{m}{0.48}, \frac{m}{0.45}, \frac{m}{0.52} \right\rangle \left\langle \frac{n}{0.54}, \frac{n}{0.55}, \frac{n}{0.46} \right\rangle\right\} \text{ and}$$

$$M_3 = \left\{\left\langle \frac{l}{0.58}, \frac{l}{0.45}, \frac{l}{0.42} \right\rangle \left\langle \frac{m}{0.64}, \frac{m}{0.45}, \frac{m}{0.36} \right\rangle \left\langle \frac{n}{0.74}, \frac{n}{0.55}, \frac{n}{0.26} \right\rangle\right\}.$$ Then (X, τ^{μ}) is a FNSTS.

Let $P_{FN} = \left\{\left\langle \frac{l}{0.34}, \frac{l}{0.45}, \frac{l}{0.66} \right\rangle \left\langle \frac{m}{0.38}, \frac{m}{0.45}, \frac{m}{0.62} \right\rangle \left\langle \frac{n}{0.44}, \frac{n}{0.55}, \frac{n}{0.56} \right\rangle\right\}$ be a FNP.

Here $P_{FN} \in L \subseteq \mathrm{N}^{*}$ and $P_{FN} \in M \subseteq \mathrm{N}^{*}$, where $\mathrm{N}^{*} = \{\mathrm{N} / \mathrm{N} \supseteq \mathrm{L} \ \& \ \mathrm{M}\}$. Then H = {L & M} is countable base for τ^{μ}. Hence (X, τ^{μ}) is FNS – SCS.

3.6. Theorem: Let (X, τ^{μ}) & (Y, δ^{μ}) are any FNSTSs and $h: X \rightarrow Y$ is onto Open and FNS – continuous mapping. If (X, τ^{μ}) is FNS – SCS then (Y, δ^{μ}) is also FNS – SCS.

Proof: Suppose (X, τ^{μ}) is a FNS – SCS. Then H is countable base for τ^{μ}, h (H) is countable collection in Y. Now show that h (H) is countable base for τ^{μ}. Let V be a FNSNHD of FNP $P_{FN}{}^{y} \in Y$. So $P_{FN}{}^{y} \in V$ and since h is onto $\exists P_{FN}{}^{x} \in X \ni h(P_{FN}{}^{x}) = P_{FN}{}^{y} \in V$. So $P_{FN}{}^{x} \in h^{-1}(V)$ is FNSNHD of (X, τ^{μ}) and it is FNS – SCS then $\exists$ $\mathrm{H}^{*} \in$ H & H $\subseteq \tau^{\mu}$ such that $P_{FN}{}^{x} \in$ H $\subseteq h^{-1}$ (V). So $h(P_{FN}{}^{x}) \in h\,(\mathrm{H}^{*}) \subseteq$ V. So $P_{FN}{}^{y} \in h(\mathrm{H}) \subseteq$ V. Hence h (H) is countable base for τ^{μ}. Therefore (Y, δ^{μ}) is also FNS – SCS.

4. FNS - Compactness in FNSTS

In this section we define two definitions of fuzzy neutrosophic supra compactness (FNS – Compact) in fuzzy neutrosophic supra topological space (FNSTS) and established several 'properties of such notions'.

4.1. Definition: Let (X, τ^{μ}) be a FNSTS. A family $\{G_k : k \in K\}$ of 'fuzzy neutrosophic supra open sets in X satisfies the condition, $\bigcup \{G_k : k \in K\} = 1_{FN}$. Then it is called fuzzy neutrosophic supra open cover (FNS – O – C) of X.

4.2. Example: Let $X = \{l, m\}$ and $\tau^{\mu} = \{0_{FN}, 1_{FN}, \gamma_1, \gamma_2, \gamma_3, \gamma_4\}$, where

$$\gamma_1 = \left\{\left\langle \frac{l}{1}, \frac{l}{1}, \frac{l}{0.15} \right\rangle \left\langle \frac{m}{1}, \frac{m}{1}, \frac{m}{0.10} \right\rangle\right\},\ \gamma_2 = \left\{\left\langle \frac{l}{0.66}, \frac{l}{0.45}, \frac{l}{0.34} \right\rangle \left\langle \frac{m}{0.74}, \frac{m}{0.45}, \frac{m}{0.26} \right\rangle\right\},$$

$$\gamma_3 = \left\{\left\langle \frac{l}{0.77}, \frac{l}{0.54}, \frac{l}{0.23} \right\rangle \left\langle \frac{m}{0.82}, \frac{m}{0.54}, \frac{m}{0.25} \right\rangle\right\} \text{and } \gamma_4 = \left\{\left\langle \frac{l}{0.85}, \frac{l}{0.55}, \frac{l}{0} \right\rangle \left\langle \frac{m}{0.95}, \frac{m}{0.55}, \frac{m}{0} \right\rangle\right\}.$$

Then (X, τ^{μ}) is a FNSTS and $\bigcup\{\gamma_k : k = 1,2,3,4\} = 1_{FN}$. Hence $\{\gamma_1, \gamma_2, \gamma_3, \gamma_4\}$ is FNS – open cover in X.

4.3. Definition: Let (X, τ^{μ}) be a FNSTS. A finite sub family of fuzzy neutrosophic supra open cover $\{G_k : k \in K\}$ on X, which is also a fuzzy neutrosophic supra open cover of X, is called fuzzy neutrosophic supra finite sub cover (FNS – F – SC) of X.

4.4. Definition: A FNSTS (X, τ^{μ}) is called FNS – Compact if each FNSO – cover of X has a FNS finite sub cover.

4.5. Example: Let $X = \{l, m\}$ and $\tau^{\mu} = \{0_{FN}, 1_{FN}, \gamma_1, \gamma_2, \gamma_3, \gamma_4\}$, where

$$\gamma_1 = \left\{\left\langle \frac{l}{1}, \frac{l}{1}, \frac{l}{0.15} \right\rangle \left\langle \frac{m}{1}, \frac{m}{1}, \frac{m}{0.10} \right\rangle\right\},\ \gamma_2 = \left\{\left\langle \frac{l}{0.66}, \frac{l}{0.45}, \frac{l}{0.34} \right\rangle \left\langle \frac{m}{0.74}, \frac{m}{0.45}, \frac{m}{0.26} \right\rangle\right\},$$

$$\gamma_3 = \left\{\left\langle \frac{l}{0.77}, \frac{l}{0.54}, \frac{l}{0.23} \right\rangle \left\langle \frac{m}{0.82}, \frac{m}{0.54}, \frac{m}{0.25} \right\rangle\right\} \text{and } \gamma_4 = \left\{\left\langle \frac{l}{0.85}, \frac{l}{0.55}, \frac{l}{0} \right\rangle \left\langle \frac{m}{0.95}, \frac{m}{0.55}, \frac{m}{0} \right\rangle\right\}.$$

Then (X, τ^{μ}) is a FNSTS and $\bigcup\{\gamma_k : k = 1,2,3,4\} = 1_{FN}$. Hence $\{\gamma_1, \gamma_2, \gamma_3, \gamma_4\}$ is FNS – open cover in X. And also $\{\gamma_1, \gamma_2, \gamma_4\}$ is FNS – F – SC of X. Therefore (X, τ^{μ}) is a FNS – compact.

4.6. Remark: Every fuzzy neutrosophic supra topological space need not be a FNS – compact following the example 4.7.

4.7. Example: Let $X = \{l, m\}$ and $\tau^{\mu} = \{0_{FN}, 1_{FN}, \gamma_1, \gamma_2, \gamma_3\}$, where

$$\gamma_1 = \left\{\left\langle \frac{l}{0.45}, \frac{l}{0.25}, \frac{l}{0.65} \right\rangle \left\langle \frac{m}{0.35}, \frac{m}{0.55}, \frac{m}{0.75} \right\rangle\right\},\ \gamma_2 = \left\{\left\langle \frac{l}{0.65}, \frac{l}{0.35}, \frac{l}{0.45} \right\rangle \left\langle \frac{m}{0.75}, \frac{m}{0.65}, \frac{m}{0.55} \right\rangle\right\} \text{ and}$$

$$\gamma_3 = \left\{\left\langle \frac{l}{0.55}, \frac{l}{0.25}, \frac{l}{0.45} \right\rangle \left\langle \frac{m}{0.45}, \frac{m}{0.55}, \frac{m}{0.65} \right\rangle\right\}.$$ Then (X, τ^{μ}) is a FNSTS but $\bigcup\{\gamma_k : k = 1,2,3\} = \gamma_2 \neq 1_{FN}$. Hence (X, τ^{μ}) is not a FNS – compact of X.

4.8. Theorem: A FNSTS (X, τ^{μ}) is FNS – compact. Let $\eta^{\mu} \subseteq \tau^{\mu}$, then (X, η^{μ}) is also FNS – compact.
Proof: Let $\{G_k : k \in K\}$ be FNS – O – C of X. Since $\eta^{\mu} \subseteq \tau^{\mu}$, $\{G_k : k \in K\}$ be FNS – O – C of X.

Since X is FNS – compact, $\exists$ FNS – F – SC $\{G_k : k \in K\}$ of X.
This turn to show that (X, η^{μ}) is also FNS – compact.

4.9. Theorem: Let (X, τ^{μ}) be a FNSTS and M, N are FNS – compact subsets of X. Then $M \cup N$ is also FNS – compact.
Proof: Let (X, τ^{μ}) be a FNSTS and M, N are FNS – compact subsets of X. To show that $M \cup N$ is also FNS – compact. Let $\{G_k : k \in K\}$ be a FNS – O – C of both M, N and $M \cup N$. Since M, N are FNS – compact subsets of X. Then M and N have FNS – F – SC. Let $\{G_k : k = 1 \text{to} K\}$ be FNS – F – SC of M and $\{G_l : l = 1 \text{to} L\}$ be FNS – F – SC of N. Then $\bigcup \{G_k : k = 1 \text{to} K\} = 1_{FN}, \bigcup \{G_l : l = 1 \text{to} L\} = 1_{FN}$.
$\bigcup \{G_k : k = 1 \text{to} K + L\} = 1_{FN}$. Hence this FNS – O – C $\{G_k : k \in K\}$ contains FNS – F – SC of $M \cup N$. Therefore $M \cup N$ is FNS – compact.

4.10. Theorem: Let (X, τ^{μ}), (Y, δ^{μ}) be FNSTSs and let $h : X \to Y$ be a FNS – continuous surjection. If (X, τ^{μ}) is FNS – compact, then (Y, δ^{μ}) is also FNS – compact.
Proof: Given that h is FNS – continuous and onto and (X, τ^{μ}) is FNS – compact.
Let us consider $\gamma_k = \{G_k : k \in K\}$ be a FNS – O – C for Y, then $\bigcup \gamma_k = 1_{FN}$.
Since h is FNS – continuous, $h^{-1}(\bigcup \gamma_k) = h^{-1}(1_{FN}) \Rightarrow \bigcup h^{-1}(\gamma_k) = 1_{FN}$.
Since γ_k is FNSO in Y, for every $k \in K$ as the map h is FNS – continuous.
Thus the family $\{h^{-1}(\gamma_k) : k \in K\}$ is a FNS – O – C for X and since X is FNS – compact.
Then $\exists$ a FNS – F – SC $\{h^{-1}(\gamma_j) : j = 1 \text{to} n\} \ni \bigcup \{h^{-1}(\gamma_j) : j = 1 \text{ to} n\} = 1_{FN}$.
Now, $h(\bigcup_{j=1}^{n} h^{-1}(\gamma_k)) = h(1_{FN}) \Rightarrow \bigcup_{j=1}^{n} h(h^{-1}(\gamma_k)) = h(1_{FN}) \Rightarrow \bigcup_{j=1}^{n} (\gamma_k) = 1_{FN}$, (as the map h is surjective). Therefore Y is FNS – compact space.

4.11. Theorem: Let (X, τ^{μ}), (Y, δ^{μ}) be FNSTSs and let $h : X \to Y$ be a FNS – open function and (Y, δ^{μ}) be a FNS – compact. Then (X, τ^{μ}) is FNS – compact.
Proof: Let $h : X \to Y$ be a FNS – open function and (Y, δ^{μ}) be a FNS – compact.
Let $\gamma_k = \{G_k : k \in K\}$ be a FNS – O – C for X. Since h is FNS – open function, $h(\gamma_k)$ is a FNS – O – C of Y. Since Y is FNS – compact, $h(\gamma_k)$ contains a FNS – F – SC, $\{h(\gamma_{k1}, \gamma_{k2}, \gamma_{k3}, \ldots\ldots \gamma_{kn})\}$. Then $\{\gamma_{k1}, \gamma_{k2}, \gamma_{k3}, \ldots\ldots \gamma_{kn}\}$ is a FNS – F – SC for X. Thus X is FNS – compact.

4.12. Theorem: The image of FNS – compact space under a FNS – continuous map is FNS – compact space.
Proof: Let $h : X \to Y$ be a FNS – continuous map from a FNS – compact space (X, τ^{μ}) onto a FNSTS (Y, δ^{μ}). Let $\gamma_k = \{G_k : k \in K\}$ be a FNS – O – C for (Y, δ^{μ}). Since h is FNS – continuous, $\{h^{-1}(\gamma_k) : k \in K\}$ is a FNS – O – C of (X, τ^{μ}). We know that (X, τ^{μ}) is FNS – compact, the FNS – O –

C $\{h^{-1}(\gamma_k):k\in K\}$ of (X,τ^{μ}) has a FNS – F – SC $\{h^{-1}(\gamma_k):k=1,2,....n\}$. Therefore $P=\bigcup_{k\in K}^{n}h^{-1}(\gamma_k)$. Then $h(p)=\bigcup_{k\in K}^{n}(\gamma_k)$, that is $Q=\bigcup_{k\in K}^{n}(\gamma_k)$. Thus $\{\gamma_{k1},\gamma_{k2},\gamma_{k3},.....\gamma_{kn}\}$ is a FNS – F – SC of $\{(\gamma_k):k\in K\}$ for (Y,δ^{μ}). Hence (Y,δ^{μ}) is a FNS – compact.

4.13. Definition: (*a*) Let (X,τ^{μ}) be a FNSTSs and *H* is FNSS on X. If a family $\{G_k:k\in K\}$ of FNSOSs on X satisfies the condition, $H\subseteq\bigcup\{G_k:k\in K\}$, then it is said to be a FNS – O – C of *H*. A finite subfamily of the FNS – O – C $\{G_k:k\in K\}$ of *H*, which is also a FNS – O – C of *H*, is said to be a FNS – F – SC of $\{G_k:k\in K\}$.

(*b*) A FNS *H* in FNSTS (X,τ^{μ}) is said to be FNS – compact iff every FNS – O – C of *H* has a FNS – F – SC.

4.14. Theorem: Let (X,τ^{μ}), (Y,δ^{μ}) be FNSTSs and let $h:X\rightarrow Y$ be a FNS – continuous function. If *H* is FNS – compact in (X,τ^{μ}) and then *h (H)* is also FNS – compact in (Y,δ^{μ}).

Proof: Let $Q=\{\gamma_k:k\in K\}$, where $\{\gamma_k=\langle\gamma_{k1},\gamma_{k2},\gamma_{k3}\rangle:k\in K\}$ be a FNS – O – C of *h (H)*.

Then, by the definition of FNS – continuity $P=\{h^{-1}(\gamma_k):k\in K\}$ is a FNS – O – C of *H*. Since *H* is FNS – compact, $\exists$ a FNS – F – SC of P, i.e., there exists $\{\gamma_{k1},\gamma_{k2},\gamma_{k3},.....\gamma_{kn}\}$ such that $H\subseteq\bigcup_{k=1}^{n}h^{-1}(\gamma_k)$. Hence $h\ (H)\subseteq h(\bigcup_{k=1}^{n}h^{-1}(\gamma_k))\subseteq\bigcup_{k=1}^{n}h(h^{-1}(\gamma_k))\subseteq\bigcup_{k=1}^{n}(\gamma_k)$. Therefore, *h (H)* is also FNS – compact.

4.15. Definition: Let (X,τ^{μ}) be a FNSTS then the family $\{G_k:k\in K\}$ of fuzzy neutrosophic supra closed sets in X satisfies finite intersection property (FIP) if every finite sub family $\{G_k:k=1,2,...,m\}$ of the family satisfies the condition, $\bigcap\{G_k:k\in K\}\neq 0_{FN}$.

4.16. Theorem: A FNSTS (X,τ^{μ}) is a FNS – compact iff the collection of FNSCSs in X having the FIP has a non empty intersection.

Proof: Suppose that X is fuzzy neutrosophic supra compact.Let $\{\gamma_k:k\in K\}$ be a family of FNSCSs in X. Also assume $\{\gamma_k:k\in K\}$ has finite intersection property. Now we have to show that $\bigcap\{\gamma_k:k\in K\}\neq 0_{FN}$. On the contrary way suppose that $\bigcap\{\gamma_k:k\in K\}=0_{FN}\Rightarrow\overline{\bigcap_{k\in K}\gamma_k}=\overline{0_{FN}}$ $\Rightarrow\bigcup_{k\in K}\overline{\gamma_k}=1_{FN}$. Clearly $\bigcup_{k\in K}\overline{\gamma_k}=\bigcup\{\overline{\gamma_k}:k\in K\}=1_{FN}$. For every $k\in K$, γ_k is FNSCS of X. Therefore $\overline{\gamma_k}$ is FNSOS of X. Thus $\{\overline{\gamma_k}:k\in K\}$ is FNS – O – C of X. Since X is FNS – compact, therefore this FNS – O – C has FNS – F – SC, say, $\bigcup\{\overline{\gamma_k}:k=1,2,....,m\}=1_{FN}$.

Then $\bigcap\{\gamma_k:k=1,2,....,m\}=0_{FN}$. Thus the above considered family does not satisfy FIP, which is a contradiction. Therefore, $\bigcap\{\gamma_k:k\in K\}\neq 0_{FN}$. Conversely, assume that the family of FNSCS of X having FIP has nonempty intersection. To show that X is fuzzy neutrosophic supra compact.

Let $\{G_k:k\in K\}$ be a FNS – O – C of X. Suppose that this FNS – O – C has no FNS – F – Sc. That is for every finite sub collection of this cover, say $\bigcup\{G_k : k \in K\} \neq 1_{FN} \Rightarrow \bigcap\{\overline{G_k} : k \in K\} \neq 0_{FN}$. As each G_k is FNSOS of X and each $\overline{G_k}$ is FNSCS of X. Thus, $\{\overline{G_k} : k \in K\}$ is a family of FNSCS of X having FIP. So by the hypothesis it has nonempty intersection, that is $\bigcap\{\overline{G_k} : k \in K\} \neq 0_{FN} \Rightarrow \bigcup\{G_k : k \in K\} \neq 1_{FN}$. This shows that $\{G_k : k \in K\}$ is not FNS – O – C for X. which is a contradiction. Therefore the given family should have a FNS – F – SC. This shows that X is fuzzy neutrosophic supra compact.

4.17. Theorem: Show that the following statements are equivalent
(I) X is FNS – compact
(II) For every $\{F_i\}$, where $F_i = \langle T_{F_i}, I_{F_i}, F_{F_i}\rangle$ of FNSC subsets of X, with $\bigcap F_i = 0_{FN}$ implies $\{F_i\}$ contains FNS – finite subclass $\{F_{i1}, F_{i2}, \ldots\ldots, F_{in}\}$ with $F_{i1} \cap F_{i2} \cap, \ldots\ldots \cap F_{in} = 0_{FN}$.
Proof: $(I) \Rightarrow (II)$. Suppose $\bigcap F_i = 0_{FN}$. Then by De Morgan's Law, $(\bigcap F_i)^C = (0_{FN})^C \Rightarrow \bigcup F_i^C = 1_{FN}$. So $\{F_i^C\}$ is a FNS – O – cover of X. Since X is FNS – compact then $\exists F_{i1}^C, F_{i2}^C, \ldots\ldots, F_{in}^C \in \{F_i^C\}$ such that $F_{i1}^C \cup F_{i2}^C \cup \ldots\ldots \cup F_{in}^C = 1_{FN}$. Then
$0_{FN} = (1_{FN})^C = (F_{i1}^C \cup F_{i2}^C \cup \ldots\ldots \cup F_{in}^C)^C = (F_{i1}^C)^C \cap (F_{i2}^C)^C \cap \ldots\ldots \cap (F_{in}^C)^C$ by De Morgan's Law $F_{i1} \cap F_{i2} \cap, \ldots\ldots \cap F_{in}$. Therefore $F_{i1} \cap F_{i2} \cap, \ldots\ldots \cap F_{in} = 0_{FN}$. So we have to prove that $(I) \Rightarrow (II)$.
$(II) \Rightarrow (I)$. Let $\{G_i\}$ be a FNS – O – C of X, where $G_i = \langle T_{G_i}, I_{G_i}, F_{G_i}\rangle$. i.e. $\bigcup G_i = 1_{FN}$.
By De Morgan's Law, $0_{FN} = (1_{FN})^C = (\bigcup G_i)^C = \bigcap G_i^C$.
Since, each G_i is FNSO, so $\{G_i^C\}$ is a class of FNSCSs & by (II). $\exists G_{i1}^C, G_{i2}^C, \ldots\ldots, G_{in}^C \in \{G_i^C\}$ Such that $G_{i1}^C \cap G_{i2}^C \cap \ldots\ldots \cap G_{in}^C = 0_{FN}$. So by De Morgan's Law,
$1_{FN} = (0_{FN})^C = (G_{i1}^C \cap G_{i2}^C \cap \ldots\ldots \cap G_{in}^C)^C = G_{i1} \cup G_{i2} \cup \ldots\ldots \cup G_{in}$. Therefore X is FNS – compact. So we have to prove that $(II) \Rightarrow (I)$.

4.18. Theorem: Let FNSTSs (X_1, τ_1^{μ}) and (X_2, τ_2^{μ}) be a FNS – compact. Then the product FNST $\tau_1^{\mu} \times \tau_2^{\mu}$ on $X_1 \times X_2$ is FNS – compact.
Proof: Consider (X_1, τ_1^{μ}) and (X_2, τ_2^{μ}) be a FNS – compact. Let $M_i = (T_{M_i}, I_{M_i}, F_{M_i}) \in \tau_1^{\mu}$ with $\bigcup M_i = 1_{FN}$ and $N_i = (T_{N_i}, I_{N_i}, F_{N_i}) \in \tau_2^{\mu}$ with $\bigcup N_i = 1_{FN}$.
Now $M_i \times N_i = (T_{M_i}, I_{M_i}, F_{M_i}) \times (T_{N_i}, I_{N_i}, F_{N_i}) = (T_{M_i} \times T_{N_i}, I_{M_i} \times I_{N_i}, F_{M_i} \times F_{N_i})$, where
$(T_{M_i} \times T_{N_i})(r,t) = \min(T_{M_i}(r), T_{N_i}(t)), (I_{M_i} \times I_{N_i})(r,t) = \min(I_{M_i}(r), I_{N_i}(t)), (F_{M_i} \times F_{N_i})(r,t) = \max(F_{M_i}(r), F_{N_i}(t))$
where $r \in X_1, t \in X_2$. So, $M_i \times N_i = 1_{FN}$. But by definition of product topology, $M_i \times N_i \in \tau_1^{\mu} \times \tau_2^{\mu}$. i.e $\{M_i \times N_i\}$ is family of FNSOSs in $X_1 \times X_2$. Choose $\bigcup(M_i \times N_i) = 1_{FN}$. Since (X_1, τ_1^{μ}) is FNS – compact then $\{M_i\}$ has FNS – finite subclass $\{M_{ij}\}$ such that $\bigcup M_{ij} = 1_{FN}$ and also (X_2, τ_2^{μ}) is FNS –

compact then $\{N_i\}$ has FNS – finite subclass $\{N_{ij}\}$ such that $\bigcup N_{ij} = 1_{FN}$, where $j = 1,2,...,n$.
Hence $\bigcup M_{ij} \times \bigcup N_{ij} = 1_{FN}$. Therefore the product FNST $\left(X_1 \times X_2, \tau_1^{\mu} \times \tau_2^{\mu}\right)$ is also FNS – compact.

5. FNS $-Q^{\mu}$ – compactness in FNSTS

In this section two definitions of FNS – Q^{μ} – compactness in fuzzy neutrosophic supra topological space and established several properties of such notions are established.

5.1. Definition: Let $\left(X, \tau^{\mu}\right)$ be a FNSTS and E be a FNSS in X. A family M $= \{P_i : i \in I\}$ of FNS sets in X, where $P_i = \left\langle T_{P_i}, I_{P_i}, F_{P_i}\right\rangle$. Then M is said to be FNS $-Q^{\mu}$ – cover of E, if $E \subseteq \bigcup P_i$,
$T_E(l) + T_{P_i}(l) \geq 1_{FN}$, for each T_{P_i} and some $l \in X$. If each P_i is FNSO then M is said to be FNS $-Q^{\mu}$ – Open cover of E.
A sub family of FNS – Q^{μ} – cover of FNSS E in X which is also FNS – Q^{μ} – cover of E is said to be FNS – Q^{μ} – sub cover of E.

5.2. Definition: A FNSS E in X is called a FNS – Q^{μ} – compact if every FNS – Q^{μ} – open cover of E has a FNS – Q^{μ} – finite sub cover. i.e. $\exists P_{i1}, P_{i2}, \ldots\ldots, P_{in} \in$ M such that $E \subseteq \bigcup P_{ij}$, $T_E(l) + T_{P_{ij}}(l) \geq 1_{FN}$, for each $T_{P_{ij}}$ and some $l \in X$, j = 1,2,.....,m.

5.3. Example: Let $X = (r,s,t)$ & $\tau^{\mu} = \{0_{FN}, 1_{FN}, P_1, P_2\}$, where

$$P_1 = \left\{\left\langle \frac{r}{0.55,0.55,0.45} \right\rangle \left\langle \frac{s}{0.65,0.55,0.35} \right\rangle \left\langle \frac{t}{0.75,0.55,0.25} \right\rangle\right\},$$

$$P_2 = \left\{\left\langle \frac{r}{0.58,0.55,0.42} \right\rangle \left\langle \frac{s}{0.68,0.55,0.32} \right\rangle \left\langle \frac{t}{0.78,0.55,0.22} \right\rangle\right\}.$$ Then $\left(X, \tau^{\mu}\right)$ is FNSTS.

Write $$E = \left\{\left\langle \frac{r}{0.48,0.55,0.52} \right\rangle \left\langle \frac{s}{0.52,0.55,0.48} \right\rangle \left\langle \frac{t}{0.63,0.55,0.37} \right\rangle\right\}$$ be a FNSS in X.

Here $E(r) \subseteq \bigcup P_i(r)$, $T_E(r) + T_{P_i}(r) \geq 1$, $E(s) \subseteq \bigcup P_i(s)$, $T_E(s) + T_{P_i}(s) \geq 1$ and $E(t) \subseteq \bigcup P_i(t)$, $T_E(t) + T_{P_i}(t) \geq 1_{FN}$. Therefore $\{P_1, P_2\}$ is a FNS – Q^{μ} – open cover of E and P_1 or P_2 is a FNS – Q^{μ} – finite sub cover of E.

5.4. Theorem: Let $\left(X, \tau^{\mu}\right)$ be a FNSTS and P, R are FNS – Q^{μ} – compact subsets of $\left(X, \tau^{\mu}\right)$. Then $P \cup R$ is FNS – Q^{μ} – compact in $\left(X, \tau^{\mu}\right)$.
Proof: Let $\left(X, \tau^{\mu}\right)$ be a FNSTS and P, R are FNS – Q^{μ} – compact subsets of $\left(X, \tau^{\mu}\right)$. To show that

$P \cup R$ is FNS – Q^{μ} – compact in (X, τ^{μ}). Let M $= \{E_i : i \in I\}$ be a FNS – Q^{μ} – open cover of P and N $= \{F_i : i \in I\}$ be a FNS – Q^{μ} – open cover of R in (X, τ^{μ}). Now $P \subseteq \bigcup E_i$ and $R \subseteq \bigcup F_i$.
$\Rightarrow P \cup R \subseteq \bigcup E_{ij} \cup \bigcup F_{ij} \Rightarrow P \cup R \subseteq \bigcup (E_{ij} \cup F_{ij})$, j = 1,2,…,m. By definition of FNS – Q^{μ} – compactness, $T_P(r) + T_{E_i}(r) \geq 1_{FN}$ and $T_R(r) + T_{F_i}(r) \geq 1_{FN}$, for some $r \in X$.
$\Rightarrow T_{P \cup R}(r) + T_{E_i \cup F_i}(r) \geq 1_{FN}$. i.e. M $\cup$ N $= \{E_i \cup F_i\}$ is FNS – Q^{μ} – cover of $P \cup R$. Since P is FNS – Q^{μ} – compact in (X, τ^{μ}). Then P has FNS – Q^{μ} – finite sub cover, $\exists P_{i1}, P_{i2}, \ldots\ldots, P_{in} \in E_i$ such that $P \subseteq \bigcup E_{ij}$, $T_P(r) + T_{E_{ij}}(r) \geq 1_{FN}$ for some $r \in X$, $j = 1,2,\ldots,m$ and also R is FNS – Q^{μ} – compact in (X, τ^{μ}). Then R has FNS – Q^{μ} – finite sub cover, $\exists R_{i1}, R_{i2}, \ldots\ldots, R_{in} \in F_i$ such that $R \subseteq \bigcup F_{ij}$, $T_R(r) + T_{F_{ij}}(r) \geq 1_{FN}$ for some $r \in X$, $j = 1,2,\ldots,m$. Now $P \subseteq \bigcup E_{ij}$ and $R \subseteq \bigcup F_{ij}$,
$\Rightarrow P \cup R \subseteq \bigcup (E_{ij} \cup F_{ij})$, Also $T_P(r) + T_{E_{ij}}(r) \geq 1_{FN}$ and $T_R(r) + T_{F_{ij}}(r) \geq 1_{FN}$, for some $r \in X$.
$\Rightarrow T_{P \cup R}(r) + T_{E_{ij} \cup F_{ij}}(r) \geq 1_{FN}$. Hence $\{E_{ij} \cup F_{ij}\}$ is a FNS – Q^{μ} – finite sub cover of $P \cup R$. Therefore $P \cup R$ is FNS – Q^{μ} – compact in (X, τ^{μ}).

5.5. Theorem: Let (X, τ^{μ}) & (Y, δ^{μ}) be FNSTSs and $h : X \to Y$ is bijective, FNSO and FNS – continuous. If FNSS E is FNS – Q^{μ} – compact in (X, τ^{μ}) and then $h(E)$ is FNS – Q^{μ} – compact in (Y, δ^{μ}).
Proof: Let N $= \{P_i \in \delta^{\mu}\}$ be a FNS – Q^{μ} – open cover of $h(E)$ with $h(E) \subseteq \bigcup P_i$ and $T_{h(E)}(r) + T_{P_{ij}}(r) \geq 1_{FN}$, for some $r \in X$. Since $P_i \in \delta^{\mu}$ then $h^{-1}(P_i) \in \tau^{\mu}$ and $h(E) \subseteq \bigcup P_i$
$\Rightarrow E \subseteq h^{-1}(\bigcup P_i)$, for some $r \in X \Rightarrow T_E(r) + T_{h^{-1}(P_{ij})}(r) \geq 1_{FN}$. Since E is FNS – Q^{μ} – compact then a family H $= \{h^{-1}(P_i) : i \in I\}$ is FNS – Q^{μ} – open cover of E. Further since E is FNS – Q^{μ} – compact in (X, τ^{μ}) then $\exists h^{-1}(P_{i1}), h^{-1}(P_{i2}), \ldots\ldots, h^{-1}(P_{in}) \in \tau^{\mu}$ such that $E \subseteq \bigcup h^{-1}(P_{ij})$ and $T_E(r) + T_{h^{-1}(P_{ij})}(r) \geq 1_{FN}$, for some $r \in X$, j = 1,2,.....,m.
$\Rightarrow h(T_E(r)) + h(T_{h^{-1}(P_{ij})})(r) \geq h(1_{FN}) \Rightarrow T_{h(E)}(r) + T_{P_{ij}}(r) \geq 1_{FN}$, as h is FNS – continuous. But $E \subseteq \bigcup h^{-1}(P_{ij}) \Rightarrow h(E) \subseteq h(\bigcup h^{-1}(P_{ij})) \Rightarrow h(E) \subseteq \bigcup P_{ij}$. Hence $P_{ij} \in \delta^{\mu}$ such that $h(E) \subseteq \bigcup P_{ij}$ & $T_{h(E)}(r) + T_{P_{ij}}(r) \geq 1_{FN}$. Therefore $h(E)$ is FNS – Q^{μ} – compact in (Y, δ^{μ}).

5.6. Theorem: Let (X, τ^{μ}) be a FNSTS and E be a FNSS on X. if a family $\{F_i : i \in I\}$ of FNSC sub sets of X with $\bigcap F_i = 0_{FN}$ implies $\{F_i\}$ contains finite subclass $\{F_{i1}, F_{i2}, \ldots\ldots, F_{in}\}$ with

$F_{i1} \cap F_{i2} \cap .... \cap F_{in} = 0_{FN}$. Then E is FNS – Q^{μ} – compact in (X, τ^{μ}).

Proof: Given that $\bigcap F_i = 0_{FN}$. Then by De Morgan's Law, $(\bigcap F_i)^C = (0_{FN})^C$.

$\Rightarrow \bigcup F_i^C = 1_{FN}$. Let M= $\{H_i : i \in I\}$ be a FNS – Q^{μ} – open cover of E in (X, τ^{μ}).

So $E \subseteq \bigcup H_i, T_E(r) + T_{H_i}(r) \geq 1_{FN}$ for some $r \in X$. Since each H_i is FNSO then $\{H_i^C\}$ is a class of FNSCSs and by given condition, $\exists H_{i1}^C, H_{i2}^C, ..., H_{in}^C \in \{H_i^C\}$ such that $H_{i1}^C \cap H_{i2}^C \cap ...... \cap H_{in}^C = 0_{FN}$.

So by De Morgan's Law, $1_{FN} = (1_{FN}^C) = (H_{i1}^C \cap H_{i2}^C \cap ...... \cap H_{in}^C)^C = H_{i1} \cup H_{i2} \cup .... \cup H_{in}$.

Hence $E \subseteq \bigcup H_{ij}, T_E(r) + T_{H_{ij}}(r) \geq 1_{FN}, j = 1,2,...., n$, for some $r \in X$. Therefore E is FNS – Q^{μ} – compact in (X, τ^{μ}).

5.7. Theorem: Let (X, τ^{μ}) be a FNSTS and M, N are FNS – Q^{μ} – compact on (X, τ^{μ}). Then $(M \times N)$ is FNS – Q^{μ} – compact in $(X \times X, \tau^{\mu} \times \tau^{\mu})$.

Proof: Let K= $\{H_i : i \in I\}$, where $H_i \in \tau^{\mu} \times \tau^{\mu}$ be a FNS – Q^{μ} – cover of $M \times N$ in $(X \times X, \tau^{\mu} \times \tau^{\mu})$. Then $M \times N \subseteq \bigcup H_i$ and $T_{M \times N}(r,t) + T_{H_i}(r,t) \geq 1_{FN}$, for some $(r,t) \in X \times X$. Now write $H_i = P_i \times R_i$, where $P_i \times R_i \in \tau^{\mu}$. Thus $M \times N \subseteq \bigcup H_i$

$\Rightarrow M \times N \subseteq \bigcup (P_i \times R_i) \Rightarrow M \subseteq \bigcup P_i, N \subseteq \bigcup R_i$. Also $T_{M \times N}(r,t) + T_{P_i \times R_i}(r,t) \geq 1_{FN}$, for some $(r,t) \in X \times X$. Hence it is clear that $T_M(r) + T_{P_i}(r) \geq 1_{FN}$, for some $r \in X$ and $T_N(t) + T_{R_i}(t) \geq 1_{FN}$, for some $t \in X$. Therefore $\{P_i : i \in I\}$ and $\{R_i : i \in I\}$ are FNS – Q^{μ} – open cover of M and N. Since M & N are compacts then $\{P_i : i \in I\}$ and $\{R_i : i \in I\}$ have FNS – Q^{μ} – finite sub covers of M and N. Say, $\{P_{ij} : j \in I_n\}$ and $\{R_{ij} : j \in I_n\}$ such that $M \subseteq \bigcup P_{ij}, T_M(r) + T_{P_{ij}}(r) \geq 1_{FN}$, for some $r \in X$ and $N \subseteq \bigcup R_{ij}, T_N(t) + T_{R_{ij}}(t) \geq 1_{FN}$, for some $t \in X$. Thus $M \times N \subseteq \bigcup (P_{ij} \times R_{ij})$. $\Rightarrow M \times N \subseteq \bigcup H_{ij}$ and $T_{M \times N}(r,t) + T_{H_{ij}}(r,t) \geq 1_{FN}$, for some $(r,t) \in X \times X$. Therefore $(M \times N)$ is also FNS – Q^{μ} – compact on $(X \times X, \tau^{\mu} \times \tau^{\mu})$.

6. Acknowledgments

The authors are highly grateful to the refers for their valuable comments and suggestions for improving the paper. The second author gratefully acknowledges the support provided by UGC, New Delhi under NFST research grant during this research work.

7. References

[1] Alblowi, S.A. Salama. A, and Mohamed Eisa, 2013, new concepts of neutrosophic sets, *inter. Jour. Math. & comp. Appli. Research,* **vol 3 (3),** pp. 95 – 102.
[2] A.S. Mashhour, A.A. Allan, F.S. Mahmoud & F.H. Khedr, on supra topological space, *Indian J. Pure*

Appl. Math. **14 (4),** 1983, 502 – 510.
[3] Bayhan, s., and D. Coker, 1996. On fuzzy separation axioms in IFTS, *BUSEFAL,* **67,** 77 – 87.
[4] C.L. Chang, fuzzy topological spaces, *J. Math. Anal. Appli.* **24** (1968), 182 – 190.
[5] D. Coker, an introduction to intuitionistic fuzzy topological spaces, *fuzzy sets & systems,* 88 (1997), **no.1,** 81 – 89.
[6] D. Coker, M. Demirci, on intuitionistic fuzzy points, *NIFS,* **1 (2),** 1995, pp. 79 – 84.
[7] D. Coker, and S. Bayhan, 2001, on separation axioms in ITS, *int. J. of math. Sci.,* **27(10),** 621-630.
[8] K.T. Atanassov, intuitionistic fuzzy sets, *fuzzy sets & systems,* 20 (1986), **no.1,** 87 – 96.
[9] L.A. Zadeh, fuzzy sets, *Inform. And control* **8**(1965), 338 – 353.
[10] M.E. Abd, EI Monsef & A.E. Ramadan, on fuzzy supra topological spaces, *Indian J. Pure Appl. Math.* **18(4),** 1987, 322 – 329.
[11] Md. Aman Mahbub, Md. Sahadat Hossain and M. Altab Hossain, some properties of compactness in IFTS, *Intern. J. Fuzzy mathematical Archive,* **vol.16, no. 2,** 2018, 39 – 48.
[12] Md. Aman Mahbub, Md. Sahadat Hossain and M. Altab Hossain, on Q – compactness in IFTS, *J. Bangladesh Acad. Sci.,* **vol. 43, No. 2,** 197 – 203, 2019.
[13] Salama, A.A. Alblowi, A.A, 2012, neutrosophic set & neutrosophic topological spaces, *ISOR jour. Of Mathematics,* **vol 3, issue 4,** pp. 31 – 35.
[14] Salama. A.A, Smarandache, Florentin & Valeri. K, 2014, neutrosophic closed set and neutrosophic continuous functions, *neutrosophic sets & systems,* **4,**pp. 4 – 8.
[15] Smarandache. F (2002), neutrosophy & neutrosophic logic. *First international conference on neutrosophic set, probability & statistics, university of New Mexico, gallup.NM, 87301, USA.*
[16] S.Das, IFTS, department of mathematics, *national institute of technology,* may, 2013.
[17] V. Amarendra Babu & J. Aswini, fuzzy neutrosophic supra topological spaces, *advances & appli. In math. Scie.* **Vol 20, issue 8,** 2021, pp. 1339 – 1347.

From Soft Point Semigroups to Soft Semigroups

Ch. Maheswari[1], **Pusuluri V N H Ravi**[2], **Nistala V.E.S. Murthy**[3] **and V. Padmavathi**[4]

[1]Department of Mathematics, Narasaraopeta Engineering College, Narasaraopet, Andhra Pradesh, INDIA
[2]Department of Mathematics, CMR Institute of Technology, Bengaluru, Karnataka, INDIA
[3]Honorary Professor, Depatment of IT & CA, AU College of Engineering, Andhra University, Visakhapatnam, Andhra Pradesh, INDIA
[4]Department of Mathematics, St.Francis de Sales College, Bengaluru, Karnataka, INDIA

E-mail: cmaheswari2014@gmail.com,hanuravi@yahoo.com, drnvesmurthy@rediffmail.com, padmavmphil@gmail.com.

Abstract. In this paper, we show that the association, the associated soft subset for a subset of a semigroup, is substructure preserving.

Key words: Soft point, Associated soft subset, Soft semigroup.

2010 Mathematics Subject Classification 20M12, 06D72.

1. Introduction

Because of the inherent relationship[1] between finite semigroups and finite automata via the syntactic monoid, semigroup theory has been of particular importance in theoretical computer science since the 1950s. Semigroups are linked to the Markov process in probability theory. Partial Differential Equations have some biological applications. Further, well-known test problems such as the decidability of dot-depth and the regularity of word classes, can be viewed as applications to semi-groups[2].

Soft set theory, introduced by Molodtsov[3], has been viewed as a novel mathematical tool for dealing with uncertainty since its beginnings, and it has a wide range of applications in the construction of algebraic structures such as groups, semirings, rings, near-rings, soft substructures, and union soft substructures. Soft set theory focuses on providing a balanced treatment of theory and practice. Approximate reasoning, , computing with words, decision support systems, expert systems, information and knowledge modelling, Intelligent systems self-adaptation and self-organizational systems have all been advocated as part of the Information Sciences discipline by Soft set theory.

However, Murthy-Maheswari[4] changed Ali-Shabir-definitions and Shum's idea of soft semigroup over a semigroup and its soft substructures. In this study, we establish the concept of soft point, which differs significantly from Sujoy Das and Samanta's definition[5]. We also show that the association, which is a soft subset for a subset of a semigroup, preserves the substructure.

2. Preliminaries

In this section, from the existing literature some basic concepts and associated notions are recalled.

Definition 2.1 *(1) A semigroup is a set with an associative binary operation.*
Let S be a semigroup and $A \subseteq S$. Then
(2) A is a subsemigroup of S iff $\forall\ x, y \in A$, $xy \in A$
(3) A is a left (right) ideal of S iff $\forall\ x \in A$ and $\forall\ y \in S$, $yx \in A$ ($xy \in A$)
(4) A is an ideal of S iff it is both a left and a right ideal of S
(5) A is a quasi-ideal of S iff $\forall\ x, y \in A$ and $\forall\ x', y' \in S$, if $z = x'x = yy'$ then $z \in A$
(6) A is a bi-ideal of S iff A is a subsemigroup of S and $\forall\ x, x' \in A$ and $\forall\ y \in S$, $xyx' \in A$.

Definition 2.2 *[3] Let S be a universal set and E be a parameter set. $\mathsf{E} = (\sigma_E, E)$ is said to be a soft set over S iff $\sigma_E : E \to P(S)$ is a map.*

Definition 2.3 *[8] The* empty soft set, *denoted by $\Phi = (\sigma_\phi, \phi)$, is a soft set with the empty parameter set.*

Definition 2.4 *[8] If $\sigma_E e = \phi\ \ \forall\ e \in E$, then E is called a* null soft set. *It is denoted by ϕ_E.*

Definition 2.5 *[7] Let A, B be two soft sets. Then A is a* soft subset *of B iff (i) $A \subseteq B$ (ii) $\sigma_A a \subseteq \sigma_B a\ \ \forall\ a \in A$. It is denoted by $\mathsf{A} \subseteq \mathsf{B}$.*

Definition 2.6 *[4] (i) Let S be a semigroup and E be a soft set over S. Then $\forall\ e \in E$, if $\sigma_E e$ is a subsemigroup of S then E is called a soft semigroup over S.*
Observe that, Φ and ϕ_E are soft semigroups over S.
(ii) Let E be a soft semigroup and A be a soft subset of E. A is a soft subsemigroup [left ideal/right ideal/ ideal/quasi-ideal/bi-ideal] of E iff $\sigma_A a$ is a subsemigroup [left ideal/right ideal/ideal/quasi-ideal/bi-ideal] of $\sigma_E a\ \forall\ a \in A$.
Observe that, both Φ and ϕ_A are soft subsemigroups [left ideals/right ideals/ideals/quasi-ideals/bi-ideals] of E.

From here onwards, we consider (1) S as a universal set and we define all the soft sets over S (2) E is a soft set means, it is a soft set over S and (3) E is a soft semigroup means, it is a soft semigroup over a semigroup S.

3. Soft Point Semigroup

In this section, the notion of associated soft subset for a subset of the set of all soft points of E is introduced and show that the association, the associated soft subset for a subset of a semigroup, is substructure preserving.

Definition 3.1 *Let E be a soft set and $\mathsf{A} \subseteq \mathsf{E}$. A is said to be a* soft point *of E, iff $\mathsf{A} = \Phi$ or $\mathsf{A} = (\sigma_A, A)$, where $A = \{a\}$ and $\sigma_A a = \{\alpha\}$, where $\alpha \in \sigma_E a$. It is denoted by (a_α).*
Let $(a_\alpha), (b_\beta)$ be two soft points of E. Then $(a_\alpha) = (b_\beta)$ iff $a = b$ and $\alpha = \beta$.
The set of all soft points of E is denoted by $\mathcal{S}^p(\mathsf{E})$, that is, $\mathcal{S}^p(\mathsf{E}) = \{(a_\alpha)/a \in E, \alpha \in \sigma_E a\} \cup \{\Phi\}$.

Definition 3.2 *Let $\mathcal{A}$ be a subset of $\mathcal{S}^p(\mathsf{E})$. Then the associated soft subset of E for $\mathcal{A}$, denoted by $\widetilde{\mathcal{A}}$, is defined by $\widetilde{\mathcal{A}} = \mathsf{A}$ iff $A = \{a \in E / \exists\ \alpha \in \sigma_E a\ \ni (a_\alpha) \in \mathcal{A}\}$ and $\sigma_A a = \cup_{(a_\alpha) \in \mathcal{A}} \{\alpha\}$.*
Observe that, (1) $\widetilde{\mathcal{A}} \subseteq \mathsf{E}$ (2) if $\mathcal{A} = \{\Phi\}$ then $\widetilde{\mathcal{A}} = \Phi$ (ii) $\widetilde{\phi} = \Phi$ (iii) $\widetilde{\mathcal{S}^p(\mathsf{E})} \subseteq \mathsf{E}$.

Definition 3.3 *Let E be a soft semigroup and $(a_\alpha), (b_\beta) \in \mathcal{S}^p(\mathsf{E})$. We define $(a_\alpha)(b_\beta)$ as*

$$(a_\alpha)(b_\beta) = \begin{cases} \Phi & \text{if } (a_\alpha) = \Phi \text{ or } (b_\beta) = \Phi \text{ or } a \neq b \\ (a_{\alpha\beta}) & \text{if } a = b \end{cases}$$

Proposition 3.4 *Let* E *be a soft semigroup. Then* $\mathcal{S}^p(\mathsf{E})$ *is a semigroup.*

Proof: Let $\mathsf{P}, \mathsf{Q}, \mathsf{R} \in \mathcal{S}^p(\mathsf{E})$.
If any one of the three is empty then we are done.
Let all the three be non-empty. Then $\mathsf{P} = (a_\alpha)$, $\mathsf{Q} = (b_\beta)$ and $\mathsf{R} = (c_\gamma)$.
Case 1: $a = b$ and $b = c$ implies $(\mathsf{PQ})\mathsf{R} = (a_{\alpha\beta\gamma}) = \mathsf{P}(\mathsf{QR})$.
Case 2: $a = b$ and $b \neq c$ implies $(\mathsf{PQ})\mathsf{R} = \Phi = \mathsf{P}(\mathsf{QR})$.
Case 3: $a \neq b$ and $b = c$ implies $(\mathsf{PQ})\mathsf{R} = \Phi = \mathsf{P}(\mathsf{QR})$.
Case 4: $a \neq b$ and $b \neq c$ implies either $c = a$ or $c \neq a$, in both the cases $(\mathsf{PQ})\mathsf{R} = \Phi = \mathsf{P}(\mathsf{QR})$.
From the above, we get that $\mathcal{S}^p(\mathsf{E})$ is a semigroup.

Proposition 3.5 *Let* $\mathcal{A}$ *be a subsemigroup of* $\mathcal{S}^p(\mathsf{E})$*. Then* $\widetilde{\mathcal{A}}$ *is a soft subsemigroup of* E*.*

Proof: If $\mathcal{A} = \phi$ or $\mathcal{A} = \{\Phi\}$, then $\widetilde{\mathcal{A}} = \Phi$ is a soft subsemigroup of E.
Let $\mathcal{A} \neq \{\Phi\}$. Then $\widetilde{\mathcal{A}} = \mathsf{A}$ implies $A = \{a \in E / \exists\, \alpha \in \sigma_E a \ni (a_\alpha) \in \mathcal{A}\}$ and $\sigma_A a = \cup_{(a_\alpha) \in \mathcal{A}}\{\alpha\}$.
By 3.2, we have $\mathsf{A} = \widetilde{\mathcal{A}} \subseteq \mathsf{E}$.
Now, we prove that A is a soft subsemigroup of E.
Let $a \in A$ be fixed and $\alpha, \beta \in \sigma_A a$.
Then $\alpha, \beta \in \sigma_E a$ as $\sigma_A a \subseteq \sigma_E a$ which impies $\alpha\beta \in \sigma_E a$ as $\sigma_E a$ is a subsemigroup of U.
$\alpha, \beta \in \sigma_A a$ implies $(a_\alpha), (a_\beta) \in \mathcal{A}$ which implies $(a_{\alpha\beta}) = (a_\alpha)(a_\beta) \in \mathcal{A}$ as $\mathcal{A}$ is a subsemigroup of $\mathcal{S}^p(\mathsf{E})$ which implies $\alpha\beta \in \sigma_A a$ which implies $\sigma_A a$ is a subsemigroup of $\sigma_E a$ or $\widetilde{\mathcal{A}} = \mathsf{A}$ is a soft subsemigroup of E.

Proposition 3.6 *Let* $\mathcal{A}$ *be a left ideal of* $\mathcal{S}^p(\mathsf{E})$*. Then* $\widetilde{\mathcal{A}}$ *is a soft (left, right) ideal of* E*.*

Proof: If $\mathcal{A} = \phi$ or $\mathcal{A} = \{\Phi\}$ then $\widetilde{\mathcal{A}} = \Phi$ is a soft left ideal of E.
Let $\mathcal{A} \neq \{\Phi\}$. Then $\widetilde{\mathcal{A}} = \mathsf{A}$ implies $A = \{a \in E / \exists\, \alpha \in \sigma_E a \ni (a_\alpha) \in \mathcal{A}\}$ and $\sigma_A a = \cup_{(a_\alpha) \in \mathcal{A}}\{\alpha\}$.
By 3.2, we have $\mathsf{A} = \widetilde{\mathcal{A}} \subseteq \mathsf{E}$.
Now we prove that A is a soft left ideal of E.
Let $a \in A$ be fixed and let $\alpha \in \sigma_A a$ and $\beta \in \sigma_E a$.
Then $\alpha \in \sigma_E a$ as $\sigma_A a \subseteq \sigma_E a$ which implies $\beta\alpha \in \sigma_E a$ as $\beta, \alpha \in \sigma_E a$ and $\sigma_E a$ is a subsemigroup of U.
$\alpha \in \sigma_A a$ and $\beta \in \sigma_E a$ implies $(a_\alpha) \in \mathcal{A}$ and $(a_\beta) \in \mathcal{S}^p(\mathsf{E})$ which implies $(a_{\beta\alpha}) = (a_\beta)(a_\alpha) \in \mathcal{A}$ as $\mathcal{A}$ is a left ideal of $\mathcal{S}^p(\mathsf{E})$ which implies $\beta\alpha \in \sigma_A a$ or $\sigma_A a$ is a left ideal of $\sigma_E a$ or A is a soft left ideal of E.

Proofs of right ideals and ideals follow the same way above.

Proposition 3.7 *Let* $\mathcal{A}$ *be a quasi ideal of* $\mathcal{S}^p(\mathsf{E})$*. Then* $\widetilde{\mathcal{A}}$ *is a soft quasi ideal of* E*.*

Proof: If $\mathcal{A} = \phi$ or $\mathcal{A} = \{\Phi\}$ then $\widetilde{\mathcal{A}} = \Phi$ is a soft quasi ideal of E.
Let $\mathcal{A} \neq \{\Phi\}$. Then $\widetilde{\mathcal{A}} = \mathsf{A}$ implies $A = \{a \in E / \exists\, \alpha \in \sigma_E a \ni (a_\alpha) \in \mathcal{A}\}$ and $\sigma_A a = \cup_{(a_\alpha) \in \mathcal{A}}\{\alpha\}$.
By 3.2, we have $\mathsf{A} = \widetilde{\mathcal{A}} \subseteq \mathsf{E}$.
Now, we show that A is a soft quasi ideal of E.
Let $a \in A$ be fixed and let $\alpha, \beta \in \sigma_A a$ and $\alpha', \beta' \in \sigma_E a$ such that $\gamma = \alpha'\alpha = \beta\beta'$.
$\alpha, \beta \in \sigma_A a$ and $\sigma_A a \subseteq \sigma_E a$ implies $\alpha, \beta \in \sigma_E a$ which implies $\alpha'\alpha, \beta\beta' \in \sigma_E a$ as $\alpha, \beta, \alpha', \beta' \in \sigma_E a$ and $\sigma_E a$ is a subsemigroup of U which implies $\gamma \in \sigma_E a$.
$\alpha, \beta \in \sigma_A a$ and $\alpha', \beta' \in \sigma_E a$ implies $(a_\alpha), (a_\beta) \in \mathcal{A}$ and $(a'_\alpha), (a'_\beta) \in \mathcal{S}^p(\mathsf{E})$.
Since $\mathcal{A}$ is a quasi ideal of $\mathcal{S}^p(\mathsf{E})$, if $(a_\gamma) = (a'_\alpha)(a_\alpha) = (a'_\beta)(a_\beta)$ then $(a_\gamma) \in \mathcal{A}$ which implies $\gamma \in \sigma_A a$ which implies $\sigma_A a$ is a quasi ideal of $\sigma_E a$ or A is a soft quasi ideal of E.

Proposition 3.8 *Let* $\mathcal{A}$ *be a bi-ideal of* $\mathcal{S}^p(\mathsf{E})$*. Then* $\widetilde{\mathcal{A}}$ *is a soft bi-ideal of* E*.*

Proof: If $\mathcal{A} = \phi$ or $\mathcal{A} = \{\Phi\}$ then $\widetilde{\mathcal{A}} = \Phi$ is a soft bi-ideal of E.
Let $\mathcal{A} \neq \{\Phi\}$. Then $\widetilde{\mathcal{A}} = \mathsf{A}$ implies $A = \{a \in E / \exists\, \alpha \in \sigma_E a \ni (a_\alpha) \in \mathcal{A}\}$ and $\sigma_A a = \cup_{(a_\alpha) \in \mathcal{A}}\{\alpha\}$.
By 3.2, we have $\mathsf{A} = \widetilde{\mathcal{A}} \subseteq \mathsf{E}$.
$\mathcal{A}$ is a bi-ideal of $\mathcal{S}^p(\mathsf{E})$ implies $\mathcal{A}$ is a subsemigroup of $\mathcal{S}^p(\mathsf{E})$ which implies by 3.5, $\mathsf{A} = \widetilde{\mathcal{A}}$ is a soft subsemigroup of E.
Now we prove that A is a soft bi-ideal of $\mathcal{S}^p(\mathsf{E})$.
Let $a \in A$ be fixed and let $\alpha, \beta \in \sigma_A a$ and $\gamma \in \sigma_E a$.
Then $\alpha, \beta \in \sigma_E a$ as $\sigma_A a \subseteq \sigma_E a$ implies $\alpha\gamma\beta \in \sigma_E a$ as $\alpha, \beta, \gamma \in \sigma_E a$ and $\sigma_E a$ is a subsemigroup of U.
$\alpha, \beta \in \sigma_A a$ and $\gamma \in \sigma_E a$ implies $(a_\alpha), (a_\beta) \in \mathcal{A}$ and $(a_\gamma) \in \mathcal{S}^p(\mathsf{E})$.
Since $\mathcal{A}$ is a bi-ideal of $\mathcal{S}^p(\mathsf{E})$, $(a_{\alpha\gamma\beta}) = (a_\alpha)(a_\gamma)(a_\beta) \in \mathcal{A}$ implies $\alpha\gamma\beta \in \sigma_A a$ or $\sigma_A a$ is a bi-ideal of $\sigma_E a$ or A is a soft bi-ideal of E.

Conclusion and Future Work: In this paper, It was shown that the association, the associated soft subset for a subset of a semigroup, is substructure preserving. In continuation of this work, during our future study, we construct a Galois connection among the set of all soft substructures of a soft semigroup and the set of all substructures of a soft point semigroup.

References

[1] Iqra and Liaqat 2021 *Journal of Mathematical Sciences & Computational Mathematics* **2** 2
[2] Jorge Almeida 1997 On a Problem of Brzozowski and Fich, *Proceedings of conference on Semi-groups and applications*,
[3] D. Molodtsov 1999 *Comput. Math. Appl.* **37** 19
[4] N.V.E.S. Murthy and Ch. Maheswari 2020 *Journal of the International Mathematical Virtual Institute* **10** 215
[5] M.I. Ali, M. Shabir and K.P. Shum 2010 *Southeast Asian Bulletin of Mathematics* **34** 595
[6] Sujoy Das and S. K. Samanta 2013 *Ann. Fuzzy Math. Inform.* **6** 77
[7] D. Pei and D. Miao 2005 *Proceedings of Granular Computing, IEEE International Conference* **2** 617
[8] M.I. Ali and M. Shabir 2010 *J. Fuzzy Math.* **18** 679
[9] N.V.E.S. Murthy and Ch. Maheswari 2017 *International Journal of Computational and Applied Mathematics* **12** 615
[10] P.A. Grillet 1995 *Semigroups: An Introduction to the Structure Theory* (New York: Marcel Dekker).

Journal of Physics: Conference Series **2332** (2022) 012006 doi:10.1088/1742-6596/2332/1/012006

Study of mathematics through indian veda's : A review

Tarunika Sharma[1], Rashi Khubnani[2], Chitiralla Subramanyam[3]
[1]Department of Mathematics, New Horizon College of Engineering Bangalore, India
[2]Department of Mathematics, New Horizon College of Engineering Bangalore, India
[3]Department of Mathematics, New Horizon College of Engineering Bangalore, India
tarunika.sharma@gmail ; rashi.khubnani@gmail.com

Abstract. Vedic Mathematics a method of conceptual calculation also reasoning. It has 16 sutras (formulas) and 13 sub-sutras(corollary). Vedic Mathematics formulas, which are mathematical concepts founded proceeding antediluvian Indian scripts called Veda meaning, knowledge reiterated by SWAMI SRI BHARATI KRISNA TIRTHAJI MAHARAJA. Due to its versatile nature and speed, it applies to many fields. This paper is an array of growth and development in the field of Vedic mathematics with a special focus on the structure of Vedic multipliers and Vedic algorithms like Urdhva Tiriyagbhyam and Nikhilam algorithms. Further an over view of Vedic Mathematics with NEP2020 is deliberated.

1.INTRODUCTION

Vedic Mathematics introduced as primeval technique of Mathematics, which was reminisced, as of the Vedas in the middle of 1911 and 1918 by Sri Bharati Krishna Tirthaji (Ifrah. G, 1998). Conferring toward his study, completely mathematics originated on sixteen Sutras or word-formulae. Vedic Mathematics is a pool of primeval wiles and methods to perform arithmetic operations swiftly and more proficiently known as Indian Mathematics named as "Vedic Mathematics" since its source out of Vedas. Specifically created commencing "Atharva Vedas" the 4th Veda. "Atharva Veda" compacts with the divisions like Engineering, Mathematics, sculpture, Medicine, also form of sciences, as we know today. Veda origins from Sanskrit term refers to 'Knowledge'. Imparting systematic mathematical footsteps, deciphering difficulties at times, which is intricate and arduous. Nevertheless, using Vedic Mathematics for Common Techniques (pertinent completely collections of certain information) and Unambiguous Methods (germane ref to unambiguous collections of particular information), numerical calculations can be completed very debauched. [1][2] Vedic Mathematics offers the solution in one line although the unadventurous technique needs numerous steps. It is an primeval method, which abridges complex numbers, squaring, multiplication, cubing, divisibility, square, and cube roots. Vedic Mathematics be able to handle even repetitive auxiliary fractions and decimals. The easiness of the Tirathji method revenues calculations is passed rationally, nevertheless. Procedures able to written down. Here are numerous profits of using a malleable, rational method. Learners able to frame own approaches; and stay imperfect to single manner. It centrals to additional inventive, engrossed, and smart learners. Problematic arithmetic complications and enormous sums can often be deciphered instantly by Tirthaji's approaches. These prominent and stunning approaches stand a share of a structure of arithmetic, which Tirthaji claims to be far more methodical than the recent method. "Vedic" Mathematics is assumed to apparent to lucid also amalgamated structure of arithmetic and its approaches remain harmonizing, undeviating, also laid-back.[3][4] This paper consists of 5 sections; section 1 explains the origin of Vedic Mathematic; section 2 discusses about the founder of Vedic Mathematics; section 3 explains about Vedic Sutras and Sub-Sutras; Section 4 discusses about the growth in the field of Vedic Mathematics which includes various previous works and the proposed work ; section 5 discusses about the Vedic Multiplier based on Urdhva Tiryakbhyam Sutra and Nikhilam Sutra ; section 6 concludes the paper.

2.ABOUT THE FOUNDER OF VEDIC MATHEMATICS

Vedic Mathematics is a method of mathematics, revealed by Indian mathematician **Jagadguru Shri Bharathi Krishna Tirthaji** in the dated in the middle of A.D. 1911 and 1918 and published his discoveries in a Vedic Mathematics Book by Tirthaji Maharaj. **Shri Bharathi Krishna Tirthaji Maharaj** was born in March 1884 in the Puri village of Orissa state. He was brilliant in subjects like mathematics, science, humanities and was outstanding in the Sanskrit language. His interests were also in spiritualism and meditation. When he was committed in meditation in the forest near Sringeri, he revived the Vedic sutras. He claims that these sutras or methods he erudite from the Vedas specifically 'Rig-Veda' unswervingly or meanderingly and he spontaneously revived them had performed meditation for 8 years. Far along, he put pen to paper the sutras on the scripts but were lost. Lastly, in the year 1957, he wrote an preliminary volume of 16 sutras, which is called Vedic Mathematics, and deliberate to transcribe other sutras later. Soon he industrialized a flume in both of his eyes and passed away in the year 1960. [5]

3.ABOUT VEDIC SUTRA'S

Vedic sutras second hand successfully for resolving factorization, HCF, divisions, squares, reciprocals, and square roots, cubes and cube roots, algebraic equations, quadratic equations, multiple simultaneous equations, biquadratic equations, , cubic equations, higher degree equations, Partial fractions, Integrations, , differential calculus, Pythagoras theorem, Apollonius Theorem, Analytical Conics and so on.[6]

3.1 *Various Vedic Sutras are—*

a) Urdhva-Tiryagbyham : Perpendicularly and crossway
b) Nikhilam Navatashcaramam Dashatah : all from 9 and the last from 10
c) Ekadhikena Purvena : By one more than the previous one
d) Paravartya Yojayet : Transfer and amend
e) Shunyam Saamyasamuccaye : When the entirety is the similar that totality is zero
f) Anurupye Shunyamanyat : If one is in relation, the other is zero
g) Sankalana-vyavakalanabhyam : By adding and by deduction
h) Puranapuranabyham : By the conclusion or non-completion
i) Chalana-Kalanabyham : Alterations and Resemblances
j) Yavadunam : Whatever the magnitude of its insufficiency
k) Vyashtisamasthi : Share and Entire
l) Shesanyankena Charamena : The remnants by the last digit
m) Sopaantyadvayamantyam : The eventual and double the last but one
n) Ekanyunena Purvena : By one fewer than the preceding one.
o) Gunitasamuchyah : The product of the entirety is equivalent to the entirety of the product.
p) Gunakasamuchyah : The influences of the entirety are equivalent to the entirety of the factors.

3.2 *Various Vedic Sub-Sutras are—*

a) Adyamadyenantyamantyena
b) Seagate Sesasamjnah
c) Anurupyena
d) Kevalaih Saptakam Gunyat
e) Vestanam

f) Shunya Anyat
g) Yavadunam Tavadunikritya Varga Yojayet
h) Antyayordashake'pi
i) Samuccayagunitah
j) Lopanasthapanabhyam
k) Vilokanam
l) Gunitasamuccayah
m) Dhvajanka
n) Dwandwa Yoga
o) Adyam Anya Madhyam

4.GROWTH OF VEDIC MATHEMATICS

4.1 Previous Work

In the year 2005 Himanshu Thapliyal and M.B Srinivas have spoken the key laborious processes in Elliptic Curve Encryption- such as point addition and doubling- by using ancient Vedic mathematics and have derive up with an well-organized elucidation for the same. [7] In the year 2009 Ramalatha, Dharani, et al., has spoken the computational difficulties in the cubing circuits by expending "Anurupya" Vedic Sutra as a key. The model outcomes demonstration that there is an huge progress in circuit recital with the use of Vedic ideologies. [8] In the year 2010 since plummeting the calculation, time in conformist Fourier Transforms -which discovers vast solicitations in the meadow of digital signal processing - the FFT algorithm were familiarized. Anvesh Kumar, Ashish Raman, et al., in his paper, have advance abridged the hardware complication in applying the FFT algorithm by accepting Vedic Mathematics, which lead to ominously dipping the number of adders and multipliers.[9] In the year 2012 Jubin Hazra has validated the usage of Vedic mathematics in the application of circular obscurity circuits and has attained about 74-97% enhancement in terms of Power-Delay-Product (PDP) as equated with the conventional circuit. In addition, Diganta Sengupta, Mahamuda Sultana, et al., have comprehensively related the speed of conformist Non-Restore type division algorithm with the Vedic division algorithm, which uses "Urdhava Trigyagbhyam" Sutra and "Nikhilam" Sutra, on the low-end processors. Their outcomes display that the conventional division takes 49.3 μs to compute the division of a 15 digit dividend, while the Vedic division does the same in 2.3μs thereby accumulative the speed by multifold! [10] In the year 2014 Shantanu Oke, Suraj Lallu, and Prathamesh Lad have prepared a relative study in the conniving of VLSI-based Distinctive Divider. SRT division, 'Newton Raphson' method, and 'Dhwajam' Sutra methods are equated and considered. The outcomes designate that the Vedic philosophies are the laidback explanations to multifaceted divisions and make the most of the system performance. In addition , a project by Anjana R, Abishna B, et al., (2014) displays the application of the Vedic multiplier by the Kogge-Stone adder. Their model outcomes authenticate that their design is the fastest multiplier and adder of the time also added misusing the benefits of Vedic mathematics. In addition Panwit Tuwanuti and Nopphagaw Thongbai have prolonged their use in multicore processors. It deals with the divide and conquers method, in which long multiplier digits are split into sub-blocks, these sub-blocks are instantaneously multiplied in diverse cores using Vedic Sutras and the outcome is gained by the concatenation of all the product terms. This technique swiftly computes the multiplication of long numbers. In view of highlighting the need for data security in communication, M Senthil Kumar and S Rajalakshmi have decorated the usage of Vedic mathematics in designing an optimal architecture for achieving mix columns process, which is an significant stage in the Advanced Encryption Principles of Cryptography. Their design evidences to be far more real than the conservative architecture. In addition, S Hemalatha and V Rajamani have validated the usage of Vedic mathematics in the design of wireless sensor networks (WSN). They defines how the usage of Vedic principles magnificently produces highly economical and improved information security systems as related with the old technique.[11] In the year 2015 Khuraijam Nelson Singh and H. Tarunkumar had read the numerous multiplier designs in VLSI. From their study, it is unblemished that the Vedic multiplier leave behind several other multipliers such as array multiplier, Wallace multiplier, Bypassing Multiplier, and Booth multiplier in numerous facets. In addition, SV Mogre and DG Bhalke have suggested the application of high-speed matrix multiplication on FPGA using the "Urdhava Trigyagbhyam" Sutra. This design indicated the important drop in Area/Speed Ratio, surges the running frequency of the multiplier, and makes the system energy efficient. Further, Aditi Tadas and Dinesh Rotate (2015. pp.317-320) have agreed a relative study of Nikhilam, Dhwajank, and Paravartya Sutras in designing a 64 Bit Divider and have examined the influence of these three Sutras in contradiction of the aspects which agree the show of the processor such as consumption of power, hardware difficulties, space application, and Speed. they have anticipated a new algorithm using Vedic ideologies, which is future more proficient and efficient as associated to the three Sutras. Later, Kunal Jadhav, Aditya Vibhute, et al., have proposed Vedic ideologies as a key to upsurge the speed of the Arithmetic Logic Unit (ALU), which was intended by employing reversibility in the circuits, to reduce power degeneracy at the cost of abridged speed. Thus imminent with power-proficient and more rapidly ALU. In the area of stirred by the working of the human brain, Anshika, Yamuna, et al., have premeditated a Vedic neuron, using the "Urdhava Trigyagbhyam" Sutra, which outpaces the typical neurons in relations of speed and energy effectiveness. The use of Vedic neurons in Artificial Neural Networks is also been presented.[12][13] In the year 2016 Anuradha Savadi and Raju Yamanshetti analyses the design of digital signal processors. It discourses the diverse algorithms used for designing high-speed DSP processors. After a complete analysis of about thirteen papers, they decided that the multiplier was designed using "Urdhava Trigyagbhyam" Sutra, and binary dividers applied using "Nikhilam navatascaramam Dasatah" and "Paravartya" sutra are more in terms of speed, area, difficulty as associated to the current approaches. Further, A study by Lavanya M and Kalaiselvi A, demonstrats the usage of Vedic ideologies in designing an Adaptive FIR filter which is used in RADAR for detecting weak RADAR signals.[14] In view of the NEP 2020, research stressed on Vedic mathematics, which is demonstrate its values. While deviousness can be swift and precise in mathematical study, it will be supportive, tortuously, for other natural sciences as well as social sciences. Specialists of Vedic mathematics say that its exercise impressively develops reasoning, analysis, and synthesis capacity, which are vital for study. Though, here are defies in learning/ teaching Vedic mathematics. Specially in the age of online learning, diverse apps or resources has to be primed. An additional challenge would be to formulate course materials/books for the undergraduate. One of the objectives of the NEP is to reinstate and re-inaugurate 'Indian-ness. Vedic mathematics can demonstrate to be an significant tool in this 'reawakening of India (Bharat)'.

4.2 Proposed work

For ingenuous high speed multipliers, the Vedic Multipliers (VMs) are selected. These multipliers offer high speed of process though they are area proficient too. The Vedic Multiplier has a irregular erection hence can be effortlessly layout in a Silicon chip. Vedic multiplier can be used for numerous usage. Since multiplication is the most essential process in numerous DSPs for execution processes like complexity, filtering, Fast Fourier Alters and in the ALU units, the Vedic Multipliers is established to be taking more speed than additional multipliers. This is due to the limited products are produced and summed up parallel. In addition, the deferral related with the Vedic Multipliers is mostly due to the breeding of carry bits through adders. In this paper Vedic multiplier is based on the Vedic Sutra's like URDHAVA and NIKHILAM :

4.2.1 *UrdhvaTiryagBhyam* - vertically and crosswise.

Ex.1. The product of 212 and 112 using above sutra is given below:

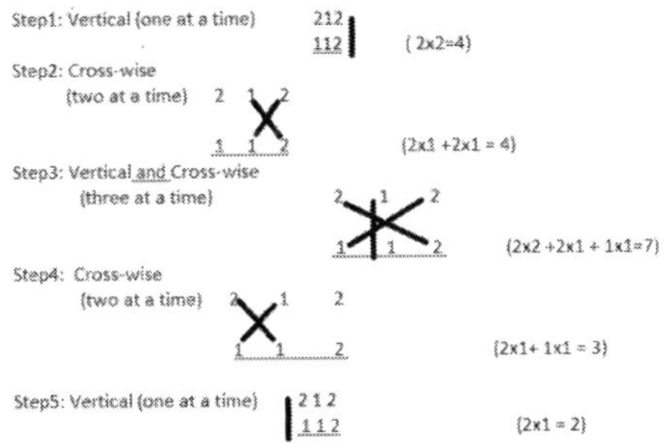

RESULT: 212 X 112 = 23744

4.2.2 *Nikhilam navatashcharamam Dashatah* - All from nine and last from ten.

Ex.2. The product of 96 and 93 using above sutra is given below:

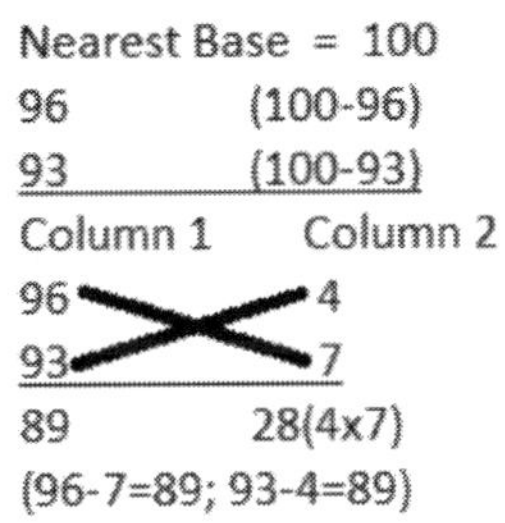

RESULT: 96 x 93 = 8928

5. VEDIC MULTIPLIER'S

5.1 Vedic Multiplier discussed in this context is based on Urdhva Tiryakbhyam Sutra:

(literally- "Vertically and Crosswise"). Using this sutra the partial products are generated and added together. This multiplier design has the advantage that as the number of bits increases, the gate delay and area increases slowly. Therefore it is time, space and area efficient. Initially the two×two Vedic Multiplier is applied with 4 AND gates and 2- 1/2 adders. By with this sutra, the superior number (N×N) is fragmented downcast addicted to smaller numbers (N/2×N/2) and this lesser numbers are again fragmented into still reduced numbers (N/4×N/4) till we grasp multiplicand of size two×two, thus abridging the entire multiplication procedure. Hence four× four Vedic Multiplier is implemented using four two×two Vedic Multipliers and three four -bit adders. Also an eight× eight Vedic Multiplier is implemented using four four×four Vedic Multipliers and three eight-bit adders. Thus an N×N Vedic Multiplier is implemented using four (N/2×N/2) Vedic Multipliers and three N-bit adders.[15][16][17][21][22][23][24][25]
The two x two Vedic multiplier segment is applied with 4 input AND gates and two half-adders which is showed in its block diagram (Fig.1)

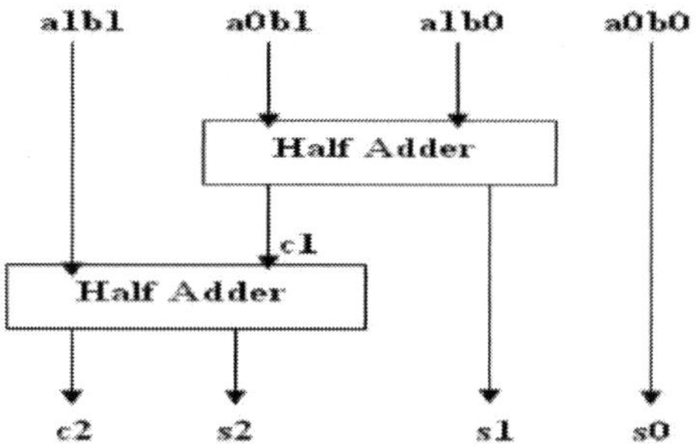

Fig.1

The 4x4 bit Vedic multiplier segment is applied using four two x two bit Vedic multiplier modules and three four-bit calculators as deliberated in Fig.3

Mockup Demonstration for four x four bit Vedic Multiplication

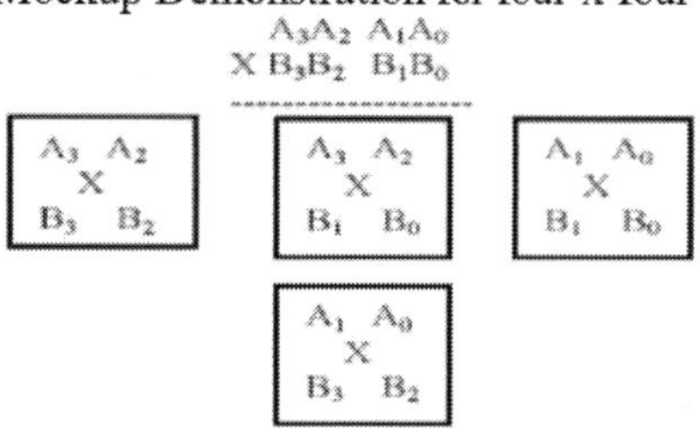

Fig.2

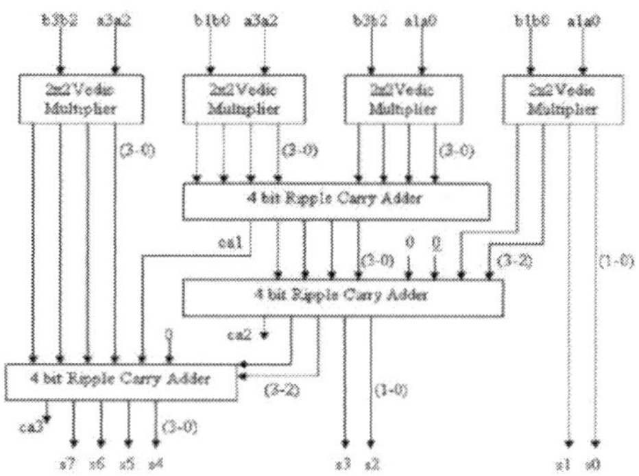

Fig.3

The eight × eight bit Vedic multiplier module is applied using four 4x4 bit Vedic multiplier modules and three eight-bit adders as discussed in Fig.4

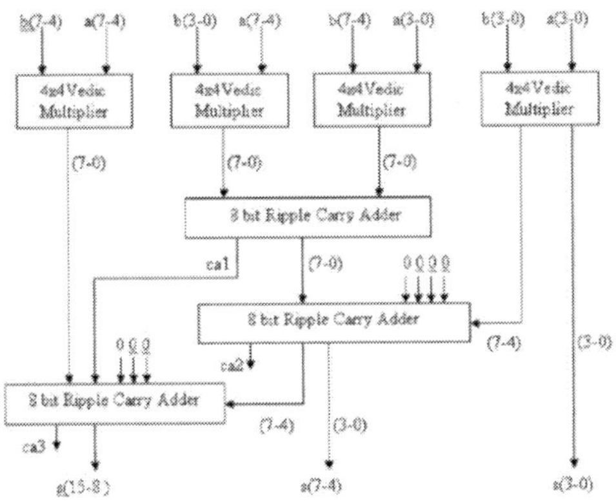

Fig.4

Indiscriminate Algorithm for $N \times N$ bit Vedic Multiplier : For AXB, the general representation for Vedic multiplication is discussed in figure 5. [18][19]

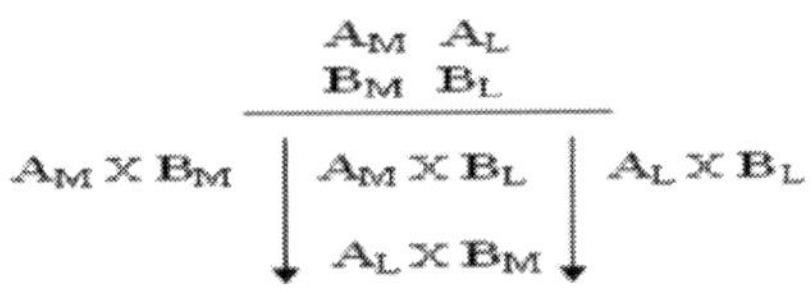

Fig.5

5.2 Vedic Multiplier discussed in this context is based on Nikhilam Sutra:

Nikhilam sutra cast-off to convert large- figure multiplication to small figure multiplication by application of following operations like subtract, add, and shift operation. A high speed multiplier can be designed by using this sutra. Also efficiently works for binary number system. It is relevant to multiplication of bases like 10, 100, 1000. The block diagram of Vedic multiplier using this sutra is shown in Fig.6 , this structure is for two inputs multiplication of 8-bit multiplier and 8-bit multiplicand operand. [20]

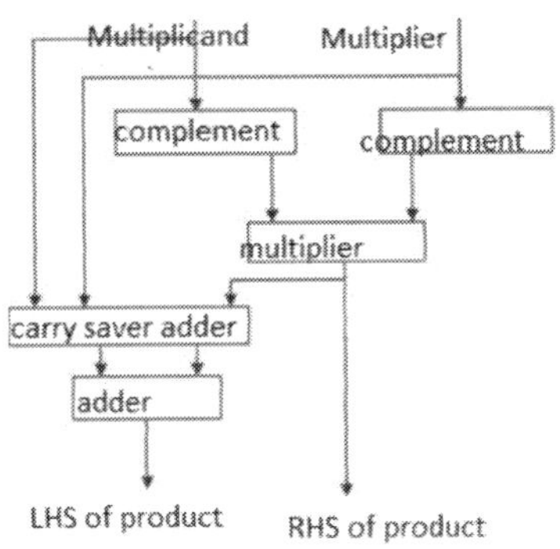

Fig.6

6. CONCLUSION

In this review paper we have discussed the advantages and fast growth of Vedic Mathematics. Since this gives faster calculation in multiplication, division, squaring, cubing etc. Therefore has widely applicable in designing Vedic Multipliers. This gives faster and better results than other multipliers. This creates a very good base for additional thinking and origination and has a huge influence on education. In view of new scheme of education NEP 2020 policy, more focus is given to Vedic Mathematics starting from primary and secondary level of education to the higher education.

7. REFERENCES

[1] Sri BharatiKrsnaTirthaji, "Vedic Mathematics", published by Motilal Banarsidass, 1965. ISBN 81- 208-0163-6.

[2] Williams K.R. "Discover Vedic Mathematics." Vedic Mathematics Research Group, 1984. ISBN 1- 869932-01-3

[3] Williams K.R. and M. Gaskell "The Cosmic Calculator". Motilal Banarsidass, 2002.ISBN 81-208- 1871-7.

[4] Nicholas A.P., Williams, J. Pickles. "Vertically and Crosswise". Inspiration Books, 1984. ISBN 1- 902517-03-2.

[5] Agrawala, V. S. (1971). Vedic Mathematics. Delhi: Motilal Banarsidas. Ancient Multiplication Methods. (n.d.). Retrieved February 17, 2010,

[6] Bibhutibhushan, D., & Avadesh, N. S. (2001). History of Hindhu Mathematics. Delhi: Bharatiya Kala Prakashan

[7] Himanshu Thapliyal, M.B Srinivas, *"An Efficient Method of Elliptic Curve Encryption Using Ancient Indian Vedic Mathematics",* IEEE 2005, pp.826-828.

[8] Ramalatha M, Thanushkodi K, Deena Dayalan K, Dharani P, *"A Novel Time and Energy Efficient Cubing Circuit using Vedic Mathematics for Finite Field Arithmetic"*, International Conference on Advances in Recent Technologies in Communication and Computing (2009).

[9] Anvesh Kumar, Ashish Raman, Dr.R.K.Sarin, Dr.Arun Khosla, *"Small area Reconfigurable FFT Design by Vedic Mathematics"*, IEEE 2010, vol. 5, pp. 836-838.

[10] Jubin Hazra, *"An Efficient Design Technique of Circular Convolution Circuit using Vedic Mathematics and McCMOS Technique",* International Conference on Computer Communication and Informatics (ICCCI -2012).

[11] Shantanu Oke, Suraj Lulla, Prathamesh Lad, *"VLSI (FPGA) Design for Distinctive Division Architecture using the Vedic Sutra 'Dhwajam'"*, 2nd IEEE International Conference on Devices, Circuits, and Systems (ICDCS), (2014).

[12] Kunal Jadhav, Aditya Vibhute, Shyam Iyer, R. Dhanabal, *"Novel Vedic Mathematics Based ALU Using Application-Specific Reversibility",* IEEE Sponsored 9th International Conference on Intelligent Systems and Control (ISCO), (2015).

[13] Anshika, Yamuna SV, Nidhi Goel, S.Indu, *"Neuronal Logic Gates Realization using Vedic Mathematics",* 1st International Conference on Next Generation Computing Technologies (NGCT-2015), pp.450-455.

[14] Anuradha Savadi, Raju Yanamshetti, *"A Survey on Design of Digital Signal Processor",* IEEE WiSPNET 2016 conference, pp. 2483 -2486, (2016).

[15] Honey Durga Tiwari, Ganzorig Gankhuyag, Chan Mo Kim and Yong Beom Cho, "Multiplier design based on ancient Indian Vedic Mathematician", International SoC Design Conference, pp. 65- 68, 2008.

[16] Parth Mehta and Dhanashri Gawali, "Conventional versus Vedic mathematics method for Hardware implementation of a multiplier", International conference on Advances in

Computing, Control, and Telecommunication Technologies, pp. 640-642, 2009.

[17] H. Thapliyal and H.R Arbania. "A Time-Area-Power Efficient Multiplier and Square Architecture Based On Ancient Indian Vedic Mathematics", Proceedings of the 2004 International Conference on VLSI (VLSI"04), Las Vegas, Nevada, June 2004, pp. 434-439.

[18] P. D. Chidgupkar and M. T. Karad, "The Implementation of Vedic Algorithms in Digital Signal Processing", Global J. of Engg. Edu, Vol.8, No.2, 2004, UICEE Published in Australia.

[19] Ch Harish Kumar. Implementation and Analysis of Power, Area and Delay of Array, Urdhva, Nikhilam Vedic Multipliers, *I. J. of Scientific and Research Publications*, Jan 2013

[20] Manoranjan Pradhan & Rutuparna Panda (2014) High speed multiplier using Nikhilam Sutra algorithm of Vedic mathematics, International Journal of Electronics, 101:3,300-307, DOI: 10.1080/00207217.2013.780298

[21] Anjana, S., C. Pradeep, and Philip Samuel."Synthesize of High Speed Floating-point Multipliers Based on Vedic Mathematics",Procedia Computer Science, 2015.

[22] Pranoti Y. Panchbhai, Nileshsingh V. Thakur, ''Performing Multiplications In Image Filtering Process Using Vedic Mathematics", International Journal of Image Processing and Vision Sciences (IJIPVS), ISSN(Print): 2278–1110,Volume-2 Issue-1

[23] Ranjana Kumari, Monali Chinchamalatpure, ''Faster Approach to Image Processing using Vedic Mathematics", IJSRD – International Journal for Scientific Research & Development| Vol.5, Issue 01, 2017 | ISSN (online):2321-0613

[24] V.M.Senthilkumar,S.Ravindrakumar,D.Nithya N.V.Kousik A vedic mathematics based Processor core for discrete wavelet transform using FinFET and CNTFET technology for biomedical signal processing, Microprocessors and Microsystems Volume 71, November 2019, 102875

[25] Udgirkar G., Indumathi G.,New Horizon college of Engineering, VLSI global routing algorithms: A survey, 2016, Proceedings of the 10th INDIACom; 2016 3rd International Conference on Computing for Sustainable Global Development, INDIACom 2016.

M𝓑J-FILTERS ON LATTICE IMPLICATION ALGEBRAS

V Amarendra Babu[1], V Siva Naga Malleswari[2], K Abida Begum[3]

[1]Assistant professor, Acharya Nagarjuna universtiy, Guntur

[2]V Siva Naga Malleswari, Assistant Professor, Prasad v potluri Siddhartha institute of technology, vijayawada.

[3]K Abida Begum, Assistant Professor, Krishna university, Machilipatnam.

[1]amarendravelisela@ymial.com, [2]vsnm.maths@gmail.com, [3]k_abida_begum@yahoo.com

Abstract. We explored the some equivalent conditions and properties of M𝓑J-filters of Lattice implication algebras. As well we define M𝓑J-lattice filters of Lattice implication algebras. Finally, it was established that M𝓑J-Lattice filter is an M𝓑J-filter, but not the other way around.

Key words: M𝓑J-filter, M𝓑J-lattice filter, Lattice implication algebra(LIA)

1.Introduction

Xu [1] proposed a logical algebra—lattice implication algebra in 1993. The lattice is defined in lattice implication algebra to explain uncertainties and particularly beyond comparison, and supposed to describe how humans reasoning. In lattice implication algebra, Xu and Qin [2] established the notion of filters and examined its properties. Xu and Qin define the fuzzy filter in lattice implication algebra and discuss some of its characteristics in [3]. Neutrosophic set was defined by Smarandache in 1998 ([4]). Neutrosophic sets are a new tool for finding uncertainty that avoids many of the problems that have plagued traditional theoretical approaches. Neutosophic set theory is being studied for a variety of applications, including information theory, probability theory, control theory, decision making, measurement theory, and so on. Since then, numerous researchers have researched in this field, and a large body of literature on the theory of neutrosophic sets has been published. The notion of M𝓑J-neutrosophic sets is introduced as another generalization of neutrosophic set.

In this paper, We explored the some equivalent conditions and properties of M𝓑J-filters of Lattice implication algebras. As well we define M𝓑J-lattice filters of Lattice implication algebras. Finally, it was established that M𝓑J-Lattice filter is an M𝓑J-filter, but not the other way around

2. Preliminaries

2.1 *Definition[8]:*

Let X be a non-empty set. A MBJ-neutrosophic set of the form $A = \{\langle \varsigma; M_A(\varsigma), \widetilde{B_A}(\varsigma), J_A(\varsigma) / \varsigma \in X\}\rangle\}$ where M_A and J_A are fuzzy sets in X, which are called a truth membership function and a false membership function, respectively and $\widetilde{B_A}$ is an IVF set in X which is called an indeterminate interval valued membership function. For the sake of simplicity, we shall use the symbol$A = (M_A, \widetilde{B_A}, J_A)$ for the M$\mathcal{B}$J-Neutrosophic set$A = \{\langle \varsigma; M_A(\varsigma), \widetilde{B_A}(\varsigma), J_A(\varsigma) / \varsigma \in X\}\rangle\}$.

In an M$\mathcal{B}$J-Neutrosophic set$A = (M_A, \widetilde{B_A}, J_A)$ in X we take $\widetilde{B_A}: X \longrightarrow [I], \varsigma \longrightarrow [B_A^-(\varsigma), B_A^+(\varsigma)]$ with $B_A^-(\varsigma) = B_A^+(\varsigma)$then $A = (M_A, \widetilde{B_A}, J_A)$ is aneutrosophic set in X.

2.2*Definition[1]*:

A bounded lattice (L, V, Λ, 0, 1) satisfy the following axioms:

(C1) $\varsigma \rightarrow (\varrho \rightarrow \varphi) = \varrho \rightarrow (\varsigma \rightarrow \varphi)$

(C2) $\varsigma \rightarrow \varsigma = 1$,

(C3) $\varsigma \rightarrow \varrho = \varrho' \rightarrow \varsigma'$,

(C4) $\varsigma \rightarrow \varrho = \varrho \rightarrow \varsigma = 1 \Rightarrow \varsigma = \varrho$,

(C5) $(\varsigma \rightarrow \varrho) \rightarrow \varrho = (\varrho \rightarrow \varsigma) \rightarrow \varsigma$,

(D1) $(\varsigma \vee \varrho) \rightarrow \varphi = (\varsigma \rightarrow \varphi) \wedge (\varrho \rightarrow \varphi)$,

(D2) $(\varsigma \wedge \varrho) \rightarrow \varphi = (\varsigma \rightarrow \varphi) \vee (\varrho \rightarrow \varphi)$, for all $\varsigma, \varrho, \varphi \in$ L.

If (L, V, Λ, 0, 1) satisfies the conditions (C1), (C2), (C3), (C4), and (C5), is called quasi-LIA.

We can define a partial ordering ≤ on a LIA L by $\varsigma \le \varrho$ if and only if $\varsigma \rightarrow \varrho = 1$.

In a LIA L, the following hold: for all $\varsigma, \varrho, \varphi \in$ L,

(1) $0 \rightarrow \varsigma = 1$, $1 \rightarrow \varsigma = \varsigma$, and $\varsigma \rightarrow 1 = 1$,

(2) $\varsigma \le \varrho \Rightarrow \varphi \rightarrow \varsigma \le \varphi \rightarrow \varrho$ and $\varsigma \rightarrow \varphi \ge \varrho \rightarrow \varphi$,

(3) $(\varsigma \rightarrow \varrho) \rightarrow ((\varrho \rightarrow \varphi) \rightarrow (\varsigma \rightarrow \varphi)) = 1$,

(4) $\varsigma \rightarrow ((\varsigma \rightarrow \varrho) \rightarrow \varrho) = 1$,

(5) $((\varsigma \rightarrow \varrho) \rightarrow \varrho) \rightarrow \varrho = \varsigma \rightarrow \varrho$.

2.3*Definition[7]*:

Let χbe the universe. A neutrosophic set $\mathbb{A}$ in χ is characterized by a truth membership function T_A, indeterminate membership function I_A and falsity membership function F_A where T_A, I_A and F_A are real standard elements of $[0,1]$. It can be written as

A= $\{\langle \varsigma, (T_A(\varsigma), I_A(\varsigma), F_A(\varsigma) / \varsigma \in E, T_A, I_A, F_A \in]0^-, 1^+[\rangle\}$. There is no restriction on the sum of $T_A(\varsigma), I_A(\varsigma)$ *and* $F_A(\varsigma)$ and so$0^- \le T_A(\varsigma) + I_A(\varsigma) + F_A(\varsigma) \le 3^+$.

3. M$\mathcal{B}$J-filters

3.1 *Definition*:

Let χbe the universe. A M $\mathcal{B}$J-neutrosophic set

$\mathbb{A} = \{\langle \varsigma; M_A(\varsigma), \widetilde{B_A}(\varsigma), J_A(\varsigma) / \varsigma \in X\}\rangle$ of a LIA L is called a M $\mathcal{B}$J-filter of L if it satisfies

(i) $M_A(1) \ge M_A(\varsigma)$, $\widetilde{B_A}(1) \ge * \widetilde{B_A}(\varsigma)$, $J_A(1) \le J_A(\varsigma)$

(ii) $M_A(\varrho) \ge \min\{M_A(\varsigma), M_A(\varsigma \rightarrow \varrho)\}$

(iii) $\widetilde{B_A}(\varrho) \ge \text{rmin}\{\widetilde{B_A}(\varsigma), \widetilde{B_A}(\varsigma \rightarrow \varrho)\}$

(iv) $J_A(\varrho) \le \max\{J_A(\varsigma), J_A(\varsigma \rightarrow \varrho)\}$

3.2 *Example*:

Let L= $\{0, a, \mathcal{b}, c, 1\}$. Define partial order on L as $0<a<\mathcal{b}<c<1$ and define $\varsigma \wedge \varrho = \min\{\varsigma, \varrho\}$ and $\varsigma \vee \varrho = \max\{\varsigma, \varrho\}$ for all $\varsigma, \varrho \in L$ and " ' "and "⟶" as follows

ς	ς'
0	1
a	c
b	b
c	a
1	0

$\rightarrow$	0	a	b	c	1
0	1	1	1	1	1
a	c	1	1	1	1
b	b	c	1	1	1
c	a	b	c	1	1
1	0	a	b	c	1

Then $(L, \vee, \wedge, ', \rightarrow)$ is LIA.
Define M𝓑J-neutrosophic set A in L by

	0	a	b	c	1
M_A	0.1	0.2	0.3	0.4	0.5
$\widetilde{B_A}$	[0.2,0.3]	[0.4,0.5]	[0.3,0.4]	[0.2,0.5]	[0.2,0.4]
J_A	0.5	0.4	0.3	0.2	0.1

Then A is a M𝓑J-filter on L.

3.3 *Theorem*:
Let L be a LIA and $A = \{\langle \varsigma; M_A(\varsigma), \widetilde{B_A}(\varsigma), J_A(\varsigma)/\varsigma \in X\}\rangle$ is a M𝓑J-filter of L, then for all$\varsigma, \varrho \in L$, $\beta \geq \varsigma$ implies $M_A(\varrho) \geq M_A(\varsigma), \widetilde{B_A}(\varrho) \geq \widetilde{B_A}(\varsigma), J_A(\varrho) \leq J_A(\varsigma)$.
Proof: Consider $M_A(\varrho) \geq \min\{M_A(\varsigma), M_A(\varsigma \rightarrow \varrho)\} \geq \min\{M_A(\varsigma), M_A(1)\} = M_A(\varsigma)$
Consider $\widetilde{B_A}(\varrho) \geq \text{rmin}\{\widetilde{B_A}(\varsigma), \widetilde{B_A}(\varsigma \rightarrow \varrho)\} \geq r\min\{\widetilde{B_A}(\varsigma), \widetilde{B_A}(1)\} = \widetilde{B_A}(\varsigma)$
Consider $J_A(\varrho) \leq \max\{J_A(\varsigma), J_A(\varsigma \rightarrow \varrho)\} \leq \max\{J_A(\varsigma), J_A(1)\} = J_A(\varsigma)$

3.4 *Theorem*:
Let L be LIA and$A = \{\langle \varsigma; M_A(\varsigma), \widetilde{B_A}(\varsigma), J_A(\varsigma)/\varsigma \in X\}\rangle$ is a M𝓑J-filter of L iff it satisfies

(i) $M_A(1) \geq M_A(\varsigma), \widetilde{B_A}(1) \geq \widetilde{B_A}(\varsigma), J_A(1) \leq J_A(\varsigma)$
(ii) $M_A(\varsigma \rightarrow \varphi) \geq \min\{M_A(\varsigma \rightarrow \varrho), M_A(\varrho \rightarrow \varphi)\}$
(iii) $\widetilde{B_A}(\varsigma \rightarrow \varphi) \geq r\min\{\widetilde{B_A}(\varsigma \rightarrow \varrho), \widetilde{B_A}(\varrho \rightarrow \varphi)\}$

(iv) $J_A(\varsigma \rightarrow \varphi) \leq \max\{J_A(\varsigma \rightarrow \varrho), J_A(\varrho \rightarrow \varphi)\}$

Proof: Assume that A is a M$\mathcal{B}$J-filter of L.
Then (i) is trivial.
To prove $M_A(\varsigma \rightarrow \varphi) \geq \min\{M_A(\varsigma \rightarrow \varrho), M_A(\varrho \rightarrow \varphi)\}$
$M_A(\varsigma \rightarrow \varphi) \geq \min\{M_A(\varrho \rightarrow \varphi), M_A((\varrho \rightarrow \varphi) \rightarrow (\varsigma \rightarrow \varphi))\}$.......(1)
Consider$(\varsigma \rightarrow \varrho) \rightarrow ((\varrho \rightarrow \varphi) \rightarrow (\varrho \rightarrow \varphi))$
$= (\varsigma \rightarrow \varrho) \rightarrow ((\varphi \rightarrow \varrho) \rightarrow (\varsigma \rightarrow \varrho))$
$=(\varphi \rightarrow \varrho) \rightarrow ((\varsigma \rightarrow \varrho) \rightarrow (\varsigma \rightarrow \varrho))$
$=(\varphi \rightarrow \varrho) \rightarrow 1$
$=1$
Then$(\varsigma \rightarrow \varrho) \leq ((\varrho \rightarrow \varphi) \rightarrow (\varsigma \rightarrow \varphi))$
$\Rightarrow M_A(\varsigma \rightarrow \varrho) \leq M_A((\varrho \rightarrow \varphi) \rightarrow (\varsigma \rightarrow \varphi))$........(2)
From (1) and (2) $M_A(\varsigma \rightarrow \varphi) \geq \min\{M_A(\varsigma \rightarrow \varrho), M_A(\varrho \rightarrow \varphi)\}$
To prove $\widetilde{\mathcal{B}_A}(\varsigma \rightarrow \varphi) \geq r\min\{\widetilde{\mathcal{B}_A}(\varsigma \rightarrow \varrho), \widetilde{\mathcal{B}_A}(\varrho \rightarrow \varphi)\}$
$\widetilde{\mathcal{B}_A}(\varsigma \rightarrow \varphi) \geq r\min\{\widetilde{\mathcal{B}_A}(\varrho \rightarrow \varphi), \widetilde{\mathcal{B}_A}((\varrho \rightarrow \varphi) \rightarrow (\varsigma \rightarrow \varphi))\}$........(3)
Since$(\varsigma \rightarrow \varrho) \leq ((\varrho \rightarrow \varphi) \rightarrow (\varsigma \rightarrow \varphi))$
$\Rightarrow \widetilde{\mathcal{B}_A}(\varsigma \rightarrow \varrho) \leq \widetilde{\mathcal{B}_A}((\varrho \rightarrow \varphi) \rightarrow (\varsigma \rightarrow \varphi))$
Therefore $\widetilde{\mathcal{B}_A}(\varsigma \rightarrow \varphi) \geq r\min\{\widetilde{\mathcal{B}_A}(\varsigma \rightarrow \varrho), \widetilde{\mathcal{B}_A}(\varrho \rightarrow \varphi)\}$
To prove $J_A(\varsigma \rightarrow \varphi) \leq \max\{J_A(\varsigma \rightarrow \varrho), J_A(\varrho \rightarrow \varphi)\}$
$J_A(\varsigma \rightarrow \varphi) \leq \max\{J_A(\varrho \rightarrow \varphi), J_A((\varrho \rightarrow \varphi) \rightarrow (\varsigma \rightarrow \varphi))\}$
Since $(\varsigma \rightarrow \varrho) \leq ((\varrho \rightarrow \varphi) \rightarrow (\varrho \rightarrow \varphi))$
$\Rightarrow J_A(\varsigma \rightarrow \varrho) \geq J_A((\varrho \rightarrow \varphi) \rightarrow (\varsigma \rightarrow \varphi))$
Therefore $J_A(\varsigma \rightarrow \varphi) \leq \max\{J_A(\varsigma \rightarrow \varrho), J_A(\varrho \rightarrow \varphi)\}$.

3.5 *Theorem:*
Let A be a Neutrosophic set and L is a LIA.
Then $A = \{(\varsigma; M_A(\varsigma), \widetilde{\mathcal{B}_A}(\varsigma), J_A(\varsigma)/\varsigma \in X\})$ is a M$\mathcal{B}$J-filter of L iff it satisfies

(i) $M_A(1) \geq M_A(\mathfrak{p})$, $\widetilde{\mathcal{B}_A}(1) \geq \widetilde{\mathcal{B}_A}(\mathfrak{p})$, $J_A(1) \leq J_A(\mathfrak{p})$
(ii) $M_A(\varphi) \geq \min\{M_A(\varsigma), M_A(\varrho), M_A(\varsigma \rightarrow (\varrho \rightarrow \varphi)\}$
(iii) $\widetilde{\mathcal{B}_A}(\varphi) \geq r\min\{\widetilde{\mathcal{B}_A}(\varsigma), \widetilde{\mathcal{B}_A}(\varrho), \widetilde{\mathcal{B}_A}(\varsigma \rightarrow (\varrho \rightarrow \varphi)\}$
(iv) $J_A(\varphi) \leq \max\{J_A(\varsigma), J_A(\varrho), J_A(\varsigma \rightarrow (\varrho \rightarrow \varphi)\}$

Proof: Suppose A is a M$\mathcal{B}$J-filter of L.
Then (i) is trivial for any $\varsigma, \varrho, \varphi \in L$.
Since $M_A(\varphi) \geq \min\{M_A(\varsigma), M_A(\varsigma \rightarrow \varphi)\}$
And $M_A(\varsigma \rightarrow \varphi) \geq \min\{M_A(\varrho), M_A(\varrho \rightarrow (\varsigma \rightarrow \varphi)\}$
Hence$M_A(\varphi) \geq \min\{M_A(\varsigma), M_A(\varrho), M_A(\varsigma \rightarrow (\varrho \rightarrow \varphi)\}$

Since$\widetilde{\mathcal{B}_A}(\varphi) \geq r\min\{\widetilde{\mathcal{B}_A}(\varsigma), \widetilde{\mathcal{B}_A}(\varsigma \rightarrow \varphi)\}$
And $\widetilde{\mathcal{B}_A}(\varsigma \rightarrow \varphi) \geq r\min\{\widetilde{\mathcal{B}_A}(\varrho), \widetilde{\mathcal{B}_A}(\varrho \rightarrow (\varsigma \rightarrow \varphi)\}$
Hence $\widetilde{\mathcal{B}_A}(\varphi) \geq r\min\{\widetilde{\mathcal{B}_A}(\varsigma), \widetilde{\mathcal{B}_A}(\varrho), \widetilde{\mathcal{B}_A}(\varsigma \rightarrow (\varrho \rightarrow \varphi)\}$
Since $J_A(\varphi) \geq max\{J_A(\varsigma), J_A(\varsigma \rightarrow \varphi)\}$
And $J_A(\varsigma \rightarrow \varphi) \geq \max\{J_A(\varrho), J_A(\varrho \rightarrow (\varsigma \rightarrow \varphi)\}$
Hence $J_A(\varphi) \geq max\{J_A(\varsigma), J_A(\varrho), J_A(\varsigma \rightarrow (\varrho \rightarrow \varphi)\}$.
Conversely
Since $M_A(\varphi) \geq \min\{M_A(\varsigma), M_A(\varrho), M_A(\varsigma \rightarrow (\varrho \rightarrow \varphi)\}$
If we take $\varsigma = \varrho$ then$M_A(\varphi) \geq \min\{M_A(\varsigma), M_A(\varsigma), M_A(\varsigma \rightarrow (\varsigma \rightarrow \varphi)\}$
$= \min\{M_A(\varsigma), M_A(\varsigma \rightarrow (\varsigma \rightarrow \varphi)\}$
Since $((\varsigma \rightarrow \varphi) \rightarrow (\varsigma \rightarrow (\varsigma \rightarrow \varphi)) = I$ then $((\varsigma \rightarrow \varphi) \leq (\varsigma \rightarrow (\varsigma \rightarrow \varphi)$
Implies $M_A(\varsigma \rightarrow \varphi) \leq M_A(\varsigma \rightarrow (\varsigma \rightarrow \varphi)$

Therefore $M_A(\varphi) \geq \min\{ M_A(\varsigma), M_A(\varsigma \rightarrow \varphi)\}$
$\widetilde{B_A}(\varphi) \geq \min\{ \widetilde{B_A}(\varsigma), \widetilde{B_A}(\varsigma \rightarrow \varphi)\}$
$J_A(\varphi) \geq max\{J_A(\varsigma), J_A(\varsigma \rightarrow \varphi)\}$.
Then A is a MBJ-filter of L.

3.6 *Theorem*:
Let A be a neutrosophic set of L. A is a MBJ-filter of L if and only if

(i) $M_A(1) \geq M_A(\varsigma)$, $\widetilde{B_A}(1) \geq \widetilde{B_A}(\varsigma)$, $J_A(1) \leq J_A(\varsigma)$
(ii) $M_A(\varsigma \rightarrow \varphi) \geq \min\{ M_A((\varphi \rightarrow \varrho) \rightarrow \varsigma), M_A(\varrho)\}$
(iii) $\widetilde{B_A}(\varsigma \rightarrow \varphi) \geq \min\{ \widetilde{B_A}((\varphi \rightarrow \varrho) \rightarrow \varsigma), \widetilde{B_A}(\varrho)\}$
(iv) $J_A(\varsigma \rightarrow \varphi) \leq \max\{J_A((\varphi \rightarrow \varrho) \rightarrow \varsigma), J_A(\varrho)\}$

for any $\mathfrak{p}, \mathfrak{q}, \varphi \in L$.
Proof: Assume that A is a MBJ-filter of L then for any $\varsigma, \varrho, \varphi \in L$
$M_A(\varphi \rightarrow \varsigma) \geq \min\{ M_A(\varrho \rightarrow (\varphi \rightarrow \varsigma)), M_A(\varrho)\}$
$\widetilde{B_A}(\varphi \rightarrow \varsigma) \geq \text{rmin}\{ \widetilde{B_A}(\varrho \rightarrow (\varphi \rightarrow \varsigma)), \widetilde{B_A}(\varrho)\}$
$J_A(\varphi \rightarrow \varsigma) \leq \max\{J_A(\varrho \rightarrow (\varphi \rightarrow \varsigma)), J_A(\varrho)\}$
Notice that for any $\varsigma, \varrho, \varphi \in L$
$((\varphi \rightarrow \varrho) \rightarrow \varsigma) \rightarrow ((\varphi \rightarrow \varrho) \rightarrow (\varphi \rightarrow \varsigma))$
$=((\varphi \rightarrow \varrho) \rightarrow \varsigma) \rightarrow (\varphi \rightarrow ((\varphi \rightarrow \varrho) \rightarrow \varsigma))$
$= \varphi \rightarrow(((\varphi \rightarrow \varrho) \rightarrow \varsigma) \rightarrow ((\varphi \rightarrow \varrho) \rightarrow \varsigma))= I$
Implies$((\varphi \rightarrow \varrho) \rightarrow \varsigma) \leq ((\varphi \rightarrow \varrho) \rightarrow (\varphi \rightarrow \varsigma))$
Implies $M_A((\varphi \rightarrow \varrho) \rightarrow \varsigma) \leq M_A((\varphi \rightarrow \varrho) \rightarrow (\varphi \rightarrow \varsigma))$
$\widetilde{B_A}((\varphi \rightarrow \varrho) \rightarrow \varsigma) \leq \widetilde{B_A}((\varphi \rightarrow \varrho) \rightarrow (\varphi \rightarrow \varsigma))$
$J_A((\varphi \rightarrow \varrho) \rightarrow \varsigma) \geq J_A((\varphi \rightarrow \varrho) \rightarrow (\varphi \rightarrow \varsigma))$
We know that $((\varphi \rightarrow (\varrho \rightarrow \varsigma)) \leq ((\varphi \rightarrow \varrho) \rightarrow (\varphi \rightarrow \varsigma))$
Implies $T_A((\varphi \rightarrow (\varrho \rightarrow \varsigma)) \leq T_A((\varphi \rightarrow \varrho) \rightarrow (\varphi \rightarrow \varsigma))$
$\widetilde{B_A}((\varphi \rightarrow (\varrho \rightarrow \varsigma)) \leq \widetilde{B_A}((\varphi \rightarrow \varrho) \rightarrow (\varphi \rightarrow \varsigma))$
$F_A((\varphi \rightarrow (\varrho \rightarrow \varsigma)) \geq F_A((\varphi \rightarrow \varrho) \rightarrow (\varphi \rightarrow \varsigma))$
Hence $M_A((\text{r}\varphi(\varrho \rightarrow \varsigma) \geq M_A((\varphi \rightarrow \varrho) \rightarrow \varsigma)$
$\widetilde{B_A}((\varphi \rightarrow (\varrho \rightarrow \varsigma)) \geq \widetilde{B_A}((\varphi \rightarrow \varrho) \rightarrow \varsigma))$
$J_A((\varphi \rightarrow (\varrho \rightarrow \varsigma)) \leq J_A((\varphi \rightarrow \varrho) \rightarrow \varsigma))$
Therefore $M_A(\varsigma \rightarrow \varphi) \geq \min\{ M_A((\varphi \rightarrow \varrho) \rightarrow \varsigma), M_A(\varrho)\}$
$\widetilde{B_A}(\varsigma \rightarrow \varphi) \geq \min\{ \widetilde{B_A}((\varphi \rightarrow \varrho) \rightarrow \varsigma), \widetilde{B_A}(\varrho)\}$
$J_A(\varsigma \rightarrow \varphi) \leq \max\{J_A((\varphi \rightarrow \varrho) \rightarrow \varsigma), J_A(\varrho)\}$
Conversely put z = I in (ii), (iii) and (iv) and from (i)

(i) Then $M_A(1) \geq M_A(\varsigma)$, $I_A(1) \geq I_A(\varsigma)$, $J_A(1) \leq J_A(\varsigma)$
(ii) $M_A(\varrho) \geq \min\{ M_A(\varsigma), M_A(\varsigma \rightarrow \varrho)\}$
(iii) $\widetilde{B_A}(\varrho) \geq r \min\{ \widetilde{B_A}(\varsigma), \widetilde{B_A}(\varsigma \rightarrow \varrho)\}$
(iv) $J_A(\varrho) \leq \max\{J_A(\varsigma), J_A(\varsigma \rightarrow \varrho)\}$

3.7 *Theorem*:
Let N be a MBJ-filter of L if $\varsigma \leq \varrho \rightarrow \varphi$ for any $\varsigma, \varrho, \varphi \in L$ then

(i) $M_A(\varphi) \geq \min\{ M_A(\varsigma), M_A(\varrho)\}$
(ii) $\widetilde{B_A}(\varphi) \geq \min\{ \widetilde{B_A}(\varsigma), \widetilde{B_A}(\varrho)\}$
(iii) $J_A(\varphi) \leq \max\{J_A(\varsigma), J_A(\varrho)\}$.

Proof: Let N be a MBJ-filter of L.
$M_A(\varphi) \geq \min\{ M_A(\varrho), M_A(\varrho \rightarrow \varphi)\}$

$\widetilde{\mathcal{B}_A}(\varphi) \geq \min\{\widetilde{\mathcal{B}_A}(\varrho), \widetilde{\mathcal{B}_A}(\varrho \longrightarrow \varphi)\}$
$J_A(\varphi) \leq \max\{J_A(\varrho), J_A(\varrho \longrightarrow \varphi)\}$
If $\varsigma \leq \varrho \longrightarrow \varphi$ then $M_A(\varsigma) \leq M_A(\varrho \longrightarrow \varphi)$
$\widetilde{\mathcal{B}_A}(\varsigma) \leq \widetilde{\mathcal{B}_A}(\varrho \longrightarrow \varphi)$
$J_A(\varsigma) \geq J_A(\varrho \longrightarrow \varphi)$
Hence $M_A(\varphi) \geq \min\{M_A(\varsigma), M_A(\varrho)\}$
$\widetilde{\mathcal{B}_A}(\varphi) \geq \min\{\widetilde{\mathcal{B}_A}(\varsigma), \widetilde{\mathcal{B}_A}(\varrho)\}$
$J_A(\varphi) \leq \max\{J_A(\varsigma), J_A(\varrho)\}$.

3.8 *Corollary*:
Let N be a M$\mathcal{B}$J-filter of L. If $(\varsigma \otimes \varrho) \longrightarrow \varphi$= I for any $\varsigma, \varrho, \varphi \in L$ then
 (i) $M_A(\varphi) \geq \min\{M_A(\varsigma), M_A(\varrho)\}$
 (ii) $\widetilde{\mathcal{B}_A}(\varphi) \geq \min\{\widetilde{\mathcal{B}_A}(\varsigma), \widetilde{\mathcal{B}_A}(\varrho)\}$
 (iii) $J_A(\varphi) \leq \max\{J_A(\varsigma), J_A(\varrho)\}$.

3.9 *Theorem*:
Let A be a neutrosophic set of L. A is a M$\mathcal{B}$J-filte of L if and only if it satisfies the following conditions for any $\varsigma, \varrho \in L$
 (i) If $\varsigma \leq \varrho \Longrightarrow M_A(\varsigma) \leq M_A(\varrho)$
 (ii) $M_A(\varsigma \otimes \varrho) \geq \min\{M_A(\varsigma), M_A(\varrho)\}$
 (iii) $\widetilde{\mathcal{B}_A}(\varsigma \otimes \varrho) \geq \min\{\widetilde{\mathcal{B}_A}(\varsigma), \widetilde{\mathcal{B}_A}(\varrho)\}$
 (iv) $J_A(\varsigma \otimes \varrho) \leq \max\{J_A(\varsigma), J_A(\varrho)\}$
Proof: Suppose N is a M$\mathcal{B}$J-filter of L.
Then (i) hold clearly for any $\varsigma, \varrho \in L$
$M_A(\varsigma \otimes \varrho) \geq \min\{M_A(\varsigma), M_A(\varsigma \longrightarrow (\varsigma \otimes \varrho))\}$
$\widetilde{\mathcal{B}_A}(\varsigma \otimes \varrho) \geq \min\{\widetilde{\mathcal{B}_A}(\varsigma), \widetilde{\mathcal{B}_A}(\varsigma \longrightarrow (\varsigma \otimes \varrho))\}$
$J_A(\varsigma \otimes \varrho) \leq \max\{J_A(\varsigma), J_A(\varsigma \longrightarrow (\varsigma \otimes \varrho))\}$
q $\longrightarrow(\varsigma \longrightarrow (\varsigma \otimes \varrho)) = \varrho \longrightarrow (\varsigma' \vee \varrho) = (\varrho \longrightarrow \varsigma_1) \vee (\varrho \longrightarrow \varrho) = (\varrho \longrightarrow \varsigma)) \vee I$=I
$(\varsigma \longrightarrow (\varsigma \otimes \varrho))) \geq \varrho$
Implies $M_A\ (\varsigma \longrightarrow (\varsigma \otimes \varrho)) \geq M_A(\varrho)$
$\widetilde{\mathcal{B}_A}\ (\varsigma \longrightarrow (\varsigma \otimes \varrho)) \geq \widetilde{\mathcal{B}_A}(\varrho)$
$J_A\ (\varsigma \longrightarrow (\varsigma \otimes \varrho)) \leq J_A(\varrho)$
Hence $M_A(\varsigma \otimes \varrho) \geq \min\{M_A(\varsigma), M_A(\varrho)\}$
$\widetilde{\mathcal{B}_A}(\varsigma \otimes \varrho) \geq \min\{\widetilde{\mathcal{B}_A}(\varsigma), \widetilde{\mathcal{B}_A}(\varrho)\}$
$J_A(\varsigma \otimes \varrho) \leq \max\{J_A(\varsigma), J_A(\varrho)\}$
Conversely if A is satisfied 1, 2, 3 and 4 then $M_A(\varsigma \otimes \varrho) \geq \min\{M_A(\varsigma), M_A(\varrho)\}$
$\widetilde{\mathcal{B}_A}(\varsigma \otimes \varrho) \geq \min\{\widetilde{\mathcal{B}_A}(\varsigma), \widetilde{\mathcal{B}_A}(\varrho)\}$
$J_A(\varsigma \otimes \varrho) \leq \max\{J_A(\varsigma), J_A(\varrho)\}$
Implies min $\{M_A(\varsigma), T_A(\varsigma \longrightarrow \varrho)\} \leq M_A(\varsigma \otimes (\varsigma \longrightarrow \varrho))$
$(\varsigma \otimes (\varsigma \longrightarrow \varrho)) \longrightarrow \varrho = (\varsigma \wedge \varrho) \longrightarrow \varrho = (\varsigma \longrightarrow \varrho) \vee (\varrho \longrightarrow \varrho)$=I
$\varrho \geq (\varsigma \otimes (\varsigma \longrightarrow \varrho))$ then $M_A(\varrho) \geq M_A\ (\varsigma \otimes (\varsigma \longrightarrow \varrho))$
It follows that min $\{M_A(\varsigma), T_A(\varsigma \longrightarrow \varrho)\} \leq M_A(\varsigma \otimes (\varsigma \longrightarrow \varrho)) \leq M_A(\varrho)$
i.e $M_A(\varrho) \geq \min\{M_A(\varsigma), M_A(\varsigma \longrightarrow \varrho)\}$
Hence A is a M$\mathcal{B}$J-filter of A

4. M $\mathcal{B}$ J-lattice filters

4.1 *Definition*:
A M $\mathcal{B}$J-Neutrosophic set A== $\{\langle \varsigma; M_A(\varsigma), \widetilde{\mathcal{B}_A}(\varsigma), J_A(\varsigma) / \varsigma \in X\}\rangle$ is on L is called a M $\mathcal{B}$ J-lattice filter if it satisfies for all $\varsigma, \varrho \in L$,

$M_A(\varsigma \wedge \varrho) \geq \min\{ M_A(\varsigma), M_A(\varrho)\}$
$\widetilde{\mathcal{B}_A}(\varsigma \wedge \varrho) \geq * \, rmin\{\widetilde{\mathcal{B}_A}(\varsigma), \widetilde{\mathcal{B}}(\varrho)\}$
$J_A(\varsigma \wedge \varrho) \leq \max\{J_A(\varsigma), J_A(\varrho)\}$

4.2 *Example*:
Let L= $\{0, a, \mathcal{b}, c, 1\}$. Define partial order on L as $0<a<\mathcal{b}<c<1$ and define $\varsigma \wedge \varrho = \min\{\varsigma, \varrho\}$ and $\varsigma \vee \varrho = \max\{\varsigma, \varrho\}$ for all $\varsigma, \varrho \in L$ and "'" and "⟶" as follows

ς	ς'
0	1
a	c
$\mathcal{b}$	$\mathcal{b}$
c	a
1	0

⟶	0	a	$\mathcal{b}$	c	1
0	1	1	1	1	1
a	c	1	1	1	1
$\mathcal{b}$	$\mathcal{b}$	c	1	1	1
c	a	$\mathcal{b}$	c	1	1
1	0	a	$\mathcal{b}$	c	1

Then (L,∨,∧,′,⟶) is LIA.
Define M $\mathcal{B}$ J-neutrosophic set A in L by

	0	a	$\mathcal{b}$	c	1
M_A	0.1	0.2	0.3	0.4	0.5
$\widetilde{\mathcal{B}_A}$	[0.2,0.3]	[0.4,0.5]	[0.3,0.4]	[0.2,0.5]	[0.2,0.4]
J_A	0.5	0.4	0.3	0.2	0.1

Then A is a M $\mathcal{B}$ J lattice filter on L.

4.3*Theorem*:
A M$\mathcal{B}$J-Neutrosophic set A=$\{\langle\varsigma; M_A(\varsigma), \widetilde{\mathcal{B}_A}(\varsigma), J_A(\varsigma)/\varsigma \in X\}\rangle$ is on L is a M$\mathcal{B}$J-filte then A is a M $\mathcal{B}$ J-lattice filter on L.
Proof: Consider $M_A(\varsigma \wedge \varrho) \geq \min\{ M_A(\varsigma), M_A((\varsigma \wedge \varrho) \longrightarrow \varrho)\}$
$= \min\{ M_A(\varsigma), M_A((\varsigma \longrightarrow \varrho) \vee (\varrho \longrightarrow \varrho)\}$
$= \min\{ M_A(\varsigma), M_A((\varsigma \longrightarrow \varrho) \vee 1\}$
$= \min\{ M_A(\varsigma), M_A(1)\}$
$\geq \min\{ M_A(\varsigma), M_A(\varrho)\}$

$\widetilde{\mathcal{B}_A}(\varsigma \wedge \varrho) \geq * \, r\min\{ \widetilde{\mathcal{B}_A}(\varsigma), \widetilde{\mathcal{B}_A}((\varsigma \wedge \varrho) \longrightarrow \varrho)\}$
$= r\min\{ \widetilde{\mathcal{B}_A}(\varsigma), \widetilde{\mathcal{B}_A}((\varsigma \longrightarrow \varrho) \vee (\varrho \longrightarrow \varrho)\}$
$= \mathrm{rmin}\{ \widetilde{\mathcal{B}_A}(\varsigma), \widetilde{\mathcal{B}_A}((\varsigma \longrightarrow \varrho) \vee 1\}$
$= r\min\{\widetilde{\mathcal{B}_A}(\varsigma), \widetilde{\mathcal{B}_A}(1)\}$
$\geq r\min\{\widetilde{\mathcal{B}_A}(\varsigma), \widetilde{\mathcal{B}_A}(\varrho)\}$

$J_A(\varsigma \wedge \varrho) \leq \max\{ J_A(\varsigma), J_A((\varsigma \wedge \varrho) \longrightarrow \varrho)\}$
$= \max\{ J_A(\varsigma), J_A((\varsigma \longrightarrow \varrho) \vee (\varrho \longrightarrow \varrho)\}$
$= \max\{ J_A(\varsigma), J_A((\varsigma \longrightarrow \varrho) \vee 1\}$
ax$\{ J_A(\varsigma), J_A(1)\}$
$\leq \max\{ J_A(\varsigma), J_A(\varrho)\}$
Then A is a M $\mathcal{B}$J-lattice filter on L.

References

[1]Y. Xu, Lattice implication algebras, Journal of Southwest Jiaotong University, 1 (1993), 20–27.
[2]Y. Xu, D. Ruan, K.Y. Qin and J. Liu, Lattice-valued logic, Studies in Fuzzyness and Soft Computing, Vol. 132, Springer-Verlag, Berlin Heidelberg, New Yorks, 2003.
[3]Y.B. Jun, E.H. Roh and Y. xu, LI-ideals in lattice implication algebras, Bulletin of the Korean Mathematical Society, 35(1) (1998), 13–24.
[4]M.A. Ozt¨urk ¨ and Y.B. Jun, Neutrosophic ideals in BCK/BCI-algebras based on neutrosophic points, Journal of the International Mathematical Virtual Institute, 8 (2018), 1–17.
[5] S.Z. Song, M. Khan, F. Smarandache and Y.B. Jun, A novel extension of neutrosophic sets and its application in BCK/BI-algebras,(2),(2018)308-318
[6] S.Z. Song, F. Smarandache and Y.B. Jun, Neutrosophic commutative N -idealsinBCK-algebras,Information,8 (2017), 130; doi:10.3390/info8040130.
[7] Y. Xu, Homomorphisms in lattice implication algebras, Proc. of 5th Many-Valued Logical Congress of China, (1992), 206–211.
[8]M.Mohseni Takallo, R.A.Borzoei, Young Bae jun, MBJ-neutrosophic structures and its applications in BCK/BCI algebras, Neuttosophic sets and systems,23(2018),72-84.

Results on Graph Energy

Sangeetha G[1] and Dr J Kavitha[2]

[1]Research scholar, Department of Applied Science, New Horizon College of Engineering, Bangalore-560103,Karnataka, India. E-mail: sangeethaganesan89@gmail.com

[2]Department of Applied Science, New Horizon College of Engineering, Bangalore-560103,Karnataka, India. E-mail: kavijossea@gmail.com

Abstract. Adjacency matrix A(G)=[aij] yields the graph energy, which is equal to the addition of absolute values of the eigenvalues G. This research investigates the energy graph class in terms of another graph class after removing a vertex. After deleting a vertex, a relationship between the energy of a complete graph E[k_n] and energy of a splitting graph E(S' [k_n]) is discovered.
Keywords: Energy of graphs, Complete graph and splitting graph.

1. Introduction

Quantum chemistry gave rise to the concept of graph energy. Erich Huckel (a German physicist) used the term "energy of graphs" in 1930 to describe the π-electron energy of a molecule of "Unsaturated conjugated hydrocarbons." Erich Huckel devised a method for determining the Schrödinger wave equation of "Unsaturated conjugated hydrocarbons" approximate solution. [1-2] contains the relevant studies.

$$E = E(G_{HMO}) = \sum_{i=1}^{n} |\lambda_i|.$$

The analogous eigenvalue is E(G_ HMO), where E(G _HMO) is the energy of the Huckel Molecular Orbital Theory (HMO) graph. In the 1970's, Ivan Gutman expanded the concept to include all graphs. Ivan realized that the following result may be applied to any graph, not just Bipartite graphs. As a result, Ivan Gutman defined graph energy in 1978.

Let G be a simple graph with E as the edge set and V as the vertex set . There are n vertices in the graph G. The Adjacency matrix A(G)=[aij] has following definition:

$$\text{The Adjacency matrix } A(G)=[a_{ij}] = \begin{cases} 1 & \text{if } v_i \text{ and } v_j \text{ are adjacent} \\ 0 & \text{otherwise.} \end{cases}$$

det(xI_n-A(G)), where I_n is the identity matrix of order n, yields the A(G) characteristic polynomial. The eigenvalues of a graph G are the same as the eigenvalues of the Adjacency matrix calculated using the characteristic polynomial [3].

In this paper, we consider a complete graph k_n $[n = 1, 2, 3, \dots]$. If $\lambda_1, \lambda_2, \lambda_3, \lambda_4, \dots, \lambda_n$ are the eigenvalues of the graph G then the graph energy E(G) is defined as follows [4].

$$E[\mathrm{G}] = \sum_{i=1}^{n} |\lambda_i|$$

Spec(G) denotes the graph's spectrum, which is determined by the graph's eigenvalues and multiplicities [8]. Numerous papers are emerged to tackle different computational hard problem [9-12].

The flow of the paper is as follows. A result on energy of complete graph is discussed in second section. In section 3, the relation between energy of splitting and complete graph is explained after deleting a vertex. Conclusion of this paper is included in last section.

2. Energy of complete graph $E[k_n]$

2.1.Result: $E[k_n] = 2\,deg(v)$. Where k_n is complete graph and v is the vertex.

Proof: By the definition of complete graph $[k_n]$ the result holds true, since the graph is complete every vertex will have equal in degree and out degree i.e., every vertex will have the same degree {n-1}.

Vertices ($n \geq 2$)	$E[k_n]$	deg($vertex$) of a complete graph	$E[k_n] = 2\deg(v)$
$n = 2$	$E[k_2] = 2$	$\deg[k_2] = 1$	$2 = 2 \times 1$
$n = 3$	$E[k_3] = 4$	$\deg[k_3] = 2$	$4 = 2 \times 2$
$n = 4$	$E[k_4] = 6$	$\deg[k_4] = 3$	$6 = 2 \times 3$
$n = 5$	$E[k_5] = 8$	$\deg[k_5] = 4$	$8 = 2 \times 4$
...	...	...	...
nth	$E[k_n]$=2{$n-1$}	$\deg[k_n] = \{n-1\}$	$E[k_n]$=2deg(v)

2.2. Result: Let $G(V,E)$ be a complete graph with m edges and n vertices, the energy of complete graph $[k_n]$ is equal to energy of a vertex deletion and twice the difference of cardinality of vertices and degree of complete graph.

$$E[k_n] = E[k_n - v] + 2[|n| - \deg(v)] \quad \text{for } (n \geq 2)$$

Proof: Consider a complete graph k_n, Every two vertices share exactly one edge. Each vertex has degree {n-1}.

We prove this result by Principle of Mathematical Induction.

We have to prove $E[k_n] = E[k_n - v] + 2[|n| - \deg(v)]$ where (n≥ 2) (1)

Basis step: n=2.

$$E[k_n] = E[k_2 - v] + 2[|2| - \deg(k_2)] \quad (\because E[k_2 = 2]$$

$$\text{Consider RHS} = E[k_2 - v] + 2[|2| - \deg(k_2)]$$

$$= E[k_1] + 2\{2 - 1\}$$

$= 0 + 2 = 2 = E[k_2]$= LHS. Hence the true for $n = 2$.

Assume that the result is true for $n = m$.

$$E[k_m] = E[k_m - v] + 2[|m| - \deg(v)] \quad \text{for } (m \geq 2) \tag{2}$$

$$E[k_m] = E[k_m - v] + 2[m - (m-1)]$$

$$E[k_m] = E[k_m - v] + 2 \tag{3}$$

The equation (3) also represent the another form of Equation (1)

Induction Step:

We have to prove the result for $n = (m + 1)$.

That is , $E[k_{m+1}] = E[k_{m+1} - v] + 2[|m+1| - \deg(v)]$ for $(m+1) \geq 2$

Consider $E[k_{m+1}] = E[k_m] + 2$ (4)

$= (E[k_m - v] + 2[|m| - \deg(v)]) + 2$

$= (E[k_{m-1}] + 2[|m| - \deg(v)]) + 2$

$= (E[k_{m-1}] + 2) + 2[|m| - \deg(v)]$

$= E[k_m] + 2[|m| - \deg(v)]$

$= E[k_{m+1} - v] + 2[|m| - \{m-1\}]$

$= E[k_{m+1} - v] + 2[|m+1| - m]$

$E[k_{m+1}] = E[k_{m+1} - v] + 2[|m+1| - \deg(v)]$.

The above equation shows that equation (1) holds true for $n = (m+1)$

Since both Basis and Induction steps are performed by Mathematical Induction, Equation (1) holds for ($n \geq 2$). Therefore the result is true $\forall\, n$, for $n \geq 2$.

Vertices ($n \geq 2$)	$E[k_n]$	$E[k_n - v]$	$2[\|n\| - \deg(v)]$	$E[k_n - v] + 2[\|n\| - \deg(v)]$
$n = 2$	$E[k_2] = 2$	$E[k_2 - v] = 0$	$2(2-1) = 2$	2
$n = 3$	$E[k_3] = 4$	$E[k_3 - v] = 2$	$2(3-2) = 2$	4
$n = 4$	$E[k_4] = 6$	$E[k_4 - v] = 4$	$2(4-3) = 2$	6
$n = 5$	$E[k_5] = 8$	$E[k_5 - v] = 6$	$2(5-4) = 2$	8
..	..	..	..	..

Corollary. $E[k_n] - E[k_{n-1}] = 2$. Difference of energy of complete graph and energy of vertex deletion is equal to 2.

2.3. Result :Difference of energy of complete graph and a vertex deletion always equal to 2

$$E[k_n] - E[k_{n-1}] = 2 \text{ for } n \geq 2.$$

Proof : The result holds true from Result (2.1)

Vertices ($n \geq 2$)	$E[k_n]$	$E[k_n - v]$
$n = 2$	$E[k_2] = 2$	$E[k_2 - v] = 0$
$n = 3$	$E[k_3] = 4$	$E[k_3 - v] = 2$
$n = 4$	$E[k_4] = 6$	$E[k_4 - v] = 4$
$n = 5$	$E[k_5] = 8$	$E[k_5 - v] = 6$
..	..	..
n^{th}	$E[k_n]$=$2\{n-1\}$	$E[k_n - v] = 2\{n-2\}$

That is, $E[k_n] - E[k_{n-1}] = 2$ for $n \geq 2$.

3. Energy of Splitting graph.

3.1. Result :$E(S'[k_n - v_1]) = \sqrt{5}\, E[k_{n-1}]$. Energy of vertex deletion on splitting of complete graph is equal to energy of k_{n-1} multiplied with $\sqrt{5}$.

3.1.1.Illustration:. Consider complete graph K_3, $[K_3 - v_1]$ is deletion of vertex from K_3 and $(S'[k_3 - v_1])$ is a splitting graph. To show that $E(S'[k_3 - v_1]) = \sqrt{5}\, E[k_{3-1}]$.

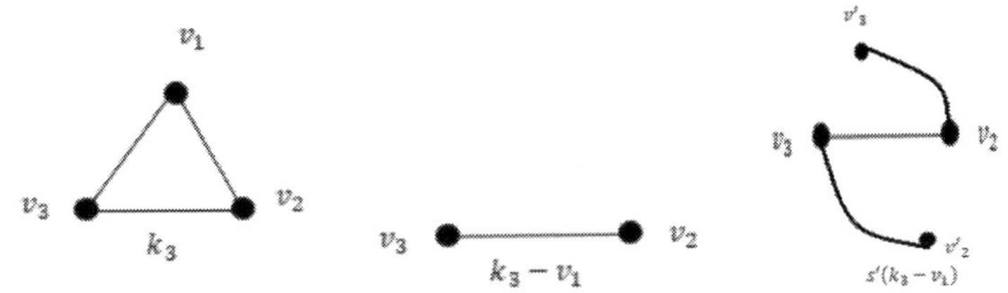

$$A(S'[k_3 - v_1]) = \begin{array}{c|cccc} & v_2 & v_3 & v'_2 & v'_3 \\ \hline v_2 & 0 & 1 & 0 & 1 \\ v_3 & 1 & 0 & 1 & 0 \\ v'_2 & 0 & 1 & 0 & 0 \\ v'_3 & 1 & 0 & 0 & 0 \end{array}$$

Characteristic Polynomial of $A(S'[k_3 - v_1])$ is $\lambda^4 - 3\lambda^2 + 1 = 0$. It's eigen values are:

$\lambda_1 = \frac{-1-\sqrt{5}}{2} = -1.6180$, $\lambda_2 = \frac{1-\sqrt{5}}{2} = -0.6180$, $\lambda_3 = \frac{-1+\sqrt{5}}{2} = 0.6180$, $\lambda_4 = \frac{1+\sqrt{5}}{2} = 1.6180$

$$Spec(S'[k_3 - v_1]) = \begin{pmatrix} \frac{-1-\sqrt{5}}{2} & \frac{1-\sqrt{5}}{2} & \frac{-1+\sqrt{5}}{2} & \frac{1+\sqrt{5}}{2} \\ 1 & 1 & 1 & 1 \end{pmatrix}$$

$$E(S'[k_3 - v_1]) = \sum_{i=1}^{n} \left| \left(\frac{-1 \pm \sqrt{5}}{2}\right)\lambda_i + \left(\frac{1 \pm \sqrt{5}}{2}\right)\lambda_i \right|$$

$=2\sqrt{5}$ can be written as

$E(S'[k_3 - v_1]) = E[k_2]\sqrt{5} = \sqrt{5}\, E[k_{n-1}]$

That is $E(S'[k_3 - v_1]) = 2\sqrt{5}$

$= E[k_2]\sqrt{5} = \sqrt{5}\, E[k_{3-1}]$

3.1.2. Illustration: Consider complete graph K_4, $[K_4 - v_1]$ is deletion of a vertex from K_4 and $(S'[k_4 - v_1])$ is a splitting graph. To show that $E(S'[k_4 - v_1]) = \sqrt{5}\, E[k_{4-1}]$.

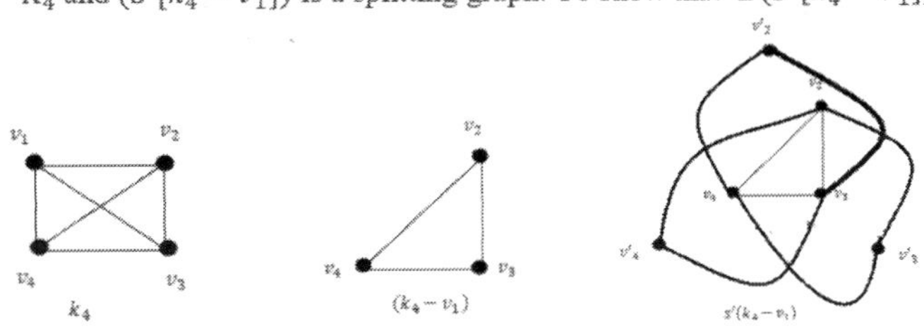

$$A(S'[k_4 - v_1]) = \begin{array}{c|cccccc} & v_2 & v_3 & v_4 & v'_2 & v'_3 & v'_4 \\ \hline v_2 & 0 & 1 & 1 & 0 & 1 & 1 \\ v_3 & 1 & 0 & 1 & 1 & 0 & 1 \\ v_4 & 1 & 1 & 0 & 1 & 1 & 0 \\ v'_2 & 0 & 1 & 1 & 0 & 0 & 0 \\ v'_3 & 1 & 0 & 1 & 0 & 0 & 0 \\ v'_4 & 1 & 1 & 0 & 0 & 0 & 0 \end{array}$$

Characteristic Polynomial of $A(S'[k_3 - v_1])$ is $\lambda^6 - 9\lambda^4 - 8\lambda^3 + 9\lambda^2 + 6\lambda^1 - 4 = 0$

It's eigenvalues are: $\lambda_1 = \frac{-1-\sqrt{5}}{2} = -1.6180$, $\lambda_2 = \frac{-1-\sqrt{5}}{2} = 1.6180$, $\lambda_3 = 1-\sqrt{5} = -1.2361$, $\lambda_4 = \frac{-1+\sqrt{5}}{2} = 0.6180$, $\lambda_5 = \frac{-1+\sqrt{5}}{2} = 0.6180$, $\lambda_6 = 1+\sqrt{5} = 3.2361$.

$$Spec\ (S'[k_4 - v_1]) = \begin{pmatrix} \frac{-1-\sqrt{5}}{2} & \frac{-1-\sqrt{5}}{2} & 1-\sqrt{5} & \frac{-1+\sqrt{5}}{2} & \frac{-1+\sqrt{5}}{2} & 1+\sqrt{5} \\ 1 & 1 & 1 & 1 & 1 & 1 \end{pmatrix}$$

$E(S'[k_4 - v_1]) = 4\sqrt{5}$ can be written as

$E(S'[k_4 - v_1]) = E[k_3]\sqrt{5} = \sqrt{5}\,E[k_{4-1}]$

$\therefore$ Energy of splitting vertex deletion on complete graph is equal to energy of k_{n-1} multiplied with $\sqrt{5}$

Thus $E(S'[k_n - v_1]) = \sqrt{5}\,E[k_{n-1}]$.

$E[k_n])$	$E(S'[k_n - v_1])$	$E(S'[k_n - v_1]) = \sqrt{5}\,E[k_{n-1}]$
$E[k_2] = 2$	$E(S'[k_2 - v_1]) = 0\sqrt{5}$	$E(S'[k_2 - v_1]) = 0\sqrt{5} = \sqrt{5}\,E[k_1]$
$E[k_3] = 4$	$E(S'[k_3 - v_1]) = 2\sqrt{5}$	$E(S'[k_3 - v_1]) = 2\sqrt{5} = \sqrt{5}\,E[k_2]$
$E[k_4] = 6$	$E(S'[k_4 - v_1]) = 4\sqrt{5}$	$E(S'[k_4 - v_1]) = 4\sqrt{5} = \sqrt{5}\,E[k_3]$
$E[k_5] = 8$	$E(S'[k_5 - v_1]) = 6\sqrt{5}$	$E(S'[k_5 - v_1]) = 6\sqrt{5} = \sqrt{5}\,E[k_4]$
...	...	...
$E[k_2] = 2$	$E(S'[k_2 - v_1]) = 0\sqrt{5}$	$E(S'[k_2 - v_1]) = 0\sqrt{5} = \sqrt{5}\,E[k_1]$

4.Conclusion

Here we have taken up the complete graph problem and its splitting graphs to investigate its energies after a vertex deletion. In this paper, the relation between energy of complete graph and its splitting graph is established for vertex deletion od a graph.

5. References

[1] C.A. Coulson, B. O'Leary, R.B. Mallion, Huckel Theory for Organic Chemists, London, 1978.

[2] K. Yates Huckel Molecular Orbital Theory, New York 1978.

[3] R. Balakrishnan, K. Ranganathan, *Textbook of Graph Theory*, Springer, New York, 2000.

[4] Gutman, *The energy of a graph*, Ber. Math. Statist. Sekt Frosch Graz 103 (1978) 1-22.

[5] Samir K. Vaidya, Kalpesh M. *Some New Results on Energy of Graphs*, ISSN 0340-6253.

[6] M.A. Sriraj, Ph.D. Thesis, *Some studies on the energy of graphs*. Mysore University, India, 2014.

[7] R. Balakrishnan, The energy of a graph, Lin. Algebra 387 (2004) 287-295.

[8] Ivan Gutman, *The energy of a graph old and new results*.

[9] Hanumesha A.G, Meenakshi K International Journal of Recent Technology and Engineering (IJRTE) ISSN: 2277-3878, **Volume-8** Issue-2S10, September 2019.

[10] Kavitha ,J, "*Star Coloring Problem*: The DNA Solution" International Journal of Information Technology and Computer Science, **Vol.4, No.3, 31-37**, 2012.

[11] Kavitha J, "*Fast Parallel DNA Algorithm based on Adleman-Lipton Model: the Independent Dominating Set Problem*", International Journal of Computer Engineering & Technology (IJCET) **Volume 6**, Issue 11, Nov 2015, pp. 01-10.

[12] Kavitha J DNA Computing towards the Solution of *Minimum Vertex Cover Problem International Journal of Psychosocial Rehabilitation*, **Vol 24**, Issue 5, ISSN: 1475-7192, page - 6807-6811, 12th May 2020 and applications, pages 196–211. Springer, 2001.

Identification of Cospectral Simple Jointed Kinematic Chains using Characteristic Polynomial of Degree Matrix and Structural Matrix

Manikandan H[1,4], Vijayananda Kaup[2], Harish Babu[3]

[1,4]Department of Mechanical Engineering (VTU-RC), CMR Institute of Technology-Bengaluru-560037, Karnataka, India
[2,3]Department of Mechanical Engineering, CMR Institute of Technology-Bengaluru-560037, Karnataka, India.

Abstract - Different methods to address the isomorphism in kinematic chains existed earlier and is well reported. In all the methods, link-adjacency matrix (LAM) represents the structural information of a given kinematic chain. Having gathered a large collection of co-spectral chains belonging to a large number of chain categories possessing (containing up to 20 links and up to 13 degrees of freedom), the present work aims at addressing the isomorphism by developing and implementing a methodology for isomorphism based on characteristic coefficients of the degree matrix and structural matrix representations of a specified Simple Jointed Kinematic Chain (SJKC) being computed and comparing. While the above two tests proved to be successful in most cases, two instances of cospectral cases were discovered wherein the above polynomial-based tests failed to detect dissimilarity in the chain structures. These cases are reported in the paper.

Keywords: Link-Adjacency matrix; Degree matrix; Structural matrix; Joint Adjacency matrix; Vertex-Incidence matrix; Isomorphism Problem; Cospectral Graphs/Chains

1. INTRODUCTION

Structural synthesis in the field of mechanisms refers to the enumeration of all kinematic chains with pre-defined number of links as well as degrees of freedom. It generally creates a collection of kinematic chains, in which, elimination of duplicate chains commonly termed as isomorphic chains has to be taken care of. To address the problem of isomorphism in a very simple yet efficient way, graph theory plays a very important role. A. H. Soni et.al [1] used graph theory in structural synthesis of kinematic chains. Uicker et.al [2] discussed a method to address the kinematic equivalence of chains where the equality of identifying numbers is used to detect the kinematic equivalency. The characteristic polynomial-based

methods were studied by Yan and Hall [3,4], Mruthyunjaya [5-7], Mruthyunjaya and Balasubramanian [8] and Sohn and Freudenstein [9]. Vijayananda [10] in his thesis used the link invariants to represent the kinematic chains and thereby addressed the isomorphism problem by generating a unique string named as isomorphism index to compare the chains. Ding et.al [11] improved and developed isomorphism research based on loop analysis of kinematic chains along with isomorphism detection as well as rigid sub-chain detection. Thus, most of the methods for isomorphism based on characteristic polynomials suffered with one or the other drawbacks or failure in some cases which resulted in co-spectral chains. Pilli et al. [12] used Eigen values and Eigen vectors for addressing isomorphism up to 10-link 1-dof chains. Manikandan et.al [16,17] used the methodology based on rooted tree structures to address isomorphism in simple jointed chains and multiple jointed kinematic chains. The generated isomorphism index was later used to identify the number of mechanisms possible from a given simple jointed kinematic chains [18]. Again, the same rooted tree structures were used to address isomorphism while synthesizing Baranov truss and 12-link & 3-d.o.f chains which established the robustness of the isomorphism index. [19,20]
The present work deals with the application of degree matrix and structural matrix for deriving the characteristic polynomial coefficients to address the isomorphism in simple jointed kinematic chains (SJKCs). The collection of cospectral cases are identified from characteristic polynomial based isomorphism test when applied to complete collection of SJKCs derived from link-adjacency matrix by same authors [13].

2. METHODOLOGY

2.1 Degree matrix

Degree matrix of a graph as defined by Mruthyunjaya et.al. [8] can be expressed as

$$[D] = \begin{cases} d_{ij} \\ 1, \ otherwise \end{cases} \tag{1.1}$$

where, d_{ij} = sum of the degrees of vertices i and j if vertex i is directly connected to vertex j;

The purpose of defining the degree matrix of a chain is to render it rich by way of packaging more amount of structural details about the chain right at the beginning. Their study has shown that several cospectral graphs, and chains containing up to 11 links were successfully identified by adopting this approach. Consider the Stephenson's chain and its equivalent graph as shown in figure 1

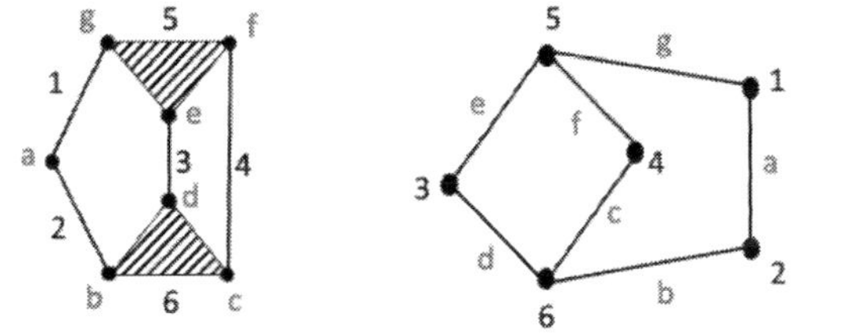

a) Stephenson's Chain b) Equivalent Graph representation

Fig. 1: Stephenson's chain and its graph representation

As per the definition of degree matrix, the degree of the links 1 to 6 is as follows:

Degree of links 1, 2, 3, 4 is 2 (as two edges emerge from vertex 1 shown in Fig. 1b). Similarly, degree of links 5 and 6 is 3.
In the degree matrix, since link1 is adjacent to link 2, the corresponding element d12 will be degree of link 1 + degree of link 2 = 4.
Following the above procedure, the [D] for the Stephenson's chain is given by,

$$[D] = \begin{bmatrix} 1 & 4 & 1 & 1 & 5 & 1 \\ 4 & 1 & 1 & 1 & 1 & 5 \\ 1 & 1 & 1 & 1 & 5 & 5 \\ 1 & 1 & 1 & 1 & 5 & 5 \\ 5 & 1 & 5 & 5 & 1 & 1 \\ 1 & 5 & 5 & 5 & 1 & 1 \end{bmatrix}$$

2.2 Structural Matrix
Similar to the degree matrix, another matrix called the 'structural matrix' was defined by Yan and Hwang [14] for kinematic chains. The effort here is to build a matrix that contains link connectivity, joint connectivity and link-joint details into one single matrix representation. Here again, the motive is to include as much structural details about the chain as possible. Due to this, the size of the matrix is larger than link-adjacency and degree matrices. The structural matrix is of order (n+m × n+m), where n and m correspond to number of links and joints present in a given kinematic chain.
If [LAM] is the (n × n) link-adjacency matrix, [IM] is the incidence matrix of the order (n × m) and [JAM] is the joint-adjacency matrix of the order (m × m), then structural matrix [S] is expressed as,

$$[S] = \begin{bmatrix} [LAM] & [IM] \\ [IM]^T & [JAM] \end{bmatrix} \quad (2.1)$$

Thus, the structural matrix contains all details about the link adjacency as well as joint adjacency of the kinematic chain. And hence, this matrix serves as an effective and exhaustive representation to calculate the characteristic coefficients as a solution to isomorphism problem in kinematic chains.
Given a kinematic chain, the elements of (n × n) link adjacency matrix [LAM], based on link-to-link connectivity is determined as follows:
Also, there cannot be a joint between same link and hence,

$$a_{ij} \begin{cases} = 1; \text{ if link } i \text{ is connected to link } j \text{ via a joint} \\ = 0; \text{ otherwise} \end{cases}$$

The link adjacency matrix for the kinematic chain shown in figure 1 is given by,

$$[LAM] = \begin{bmatrix} 0 & 1 & 0 & 0 & 1 & 0 \\ 1 & 0 & 0 & 0 & 0 & 1 \\ 0 & 0 & 0 & 0 & 1 & 1 \\ 0 & 0 & 0 & 0 & 1 & 1 \\ 1 & 0 & 1 & 1 & 0 & 0 \\ 0 & 1 & 1 & 1 & 0 & 0 \end{bmatrix}$$

The incidence matrix [IM] of order (n × m) for a given kinematic chain can be defined as follows.

Suppose a kinematic chain contains n number of links and m number of joints and let 'i' be represented as the link number and 'j' be represented as the joint number associated with the link-i, then, the incidence matrix [IM] is given by the expression,

$$b_{ij} \begin{cases} = 1;\ \text{if link i is associated with joint j} \\ = 0;\ \text{otherwise} \end{cases}$$

The incidence matrix for the Stephenson's chain shown in figure 1 can be obtained as follows.
Let the rows are designated by the link numbers i from 1 to 6 and the columns j are designated by the joint numbers from a to g (Refer figure 1b).
The joints associated to link1 are joint aand joint g. Thus in [IM]elements (1,a) and (1,g) is assigned '1' and all other elements in the first row '0'. This is followed for all the remaining rows. The resultant incidence-matrix for the Stephenson's chain is as shown below.

$$[IM] = \begin{bmatrix} 1 & 0 & 0 & 0 & 0 & 0 & 1 \\ 1 & 1 & 0 & 0 & 0 & 0 & 0 \\ 0 & 0 & 0 & 1 & 1 & 0 & 0 \\ 0 & 0 & 1 & 0 & 0 & 1 & 0 \\ 0 & 0 & 0 & 1 & 1 & 1 & 1 \\ 0 & 0 & 1 & 0 & 0 & 0 & 0 \end{bmatrix}$$

The order of incidence matrix is (n × m) viz., (6 × 7) in the example shown above. Also, it is evident that the incidence matrix is not a square symmetric matrix like link adjacency matrix.
Joint adjacency matrix of order (m × m) can be defined as follows.
Suppose a given kinematic chain contains m number of joints, then joint adjacency matrix can be derived based on the joint-joint connectivity of the given kinematic chain as expressed below.

$$c_{ij} \begin{cases} = 1;\ \text{if joint } i \text{ is associated with joint } j \text{ via a link} \\ = 0;\ \text{otherwise} \end{cases}$$

Therefore, the joint adjacency matrix for the Stephenson chain shown in figure 1 is given by,

$$[JAM] = \begin{bmatrix} 0 & 1 & 0 & 0 & 0 & 0 & 1 \\ 1 & 0 & 0 & 1 & 0 & 0 & 0 \\ 0 & 0 & 0 & 1 & 0 & 1 & 0 \\ 0 & 1 & 1 & 0 & 1 & 0 & 0 \\ 0 & 0 & 0 & 1 & 0 & 1 & 1 \\ 0 & 0 & 1 & 0 & 1 & 0 & 1 \\ 1 & 0 & 0 & 0 & 1 & 1 & 0 \end{bmatrix}$$

Having been determined the link-adjacency matrix [LAM], Incidence matrix [IM] and joint-adjacency matrix [JAM], the structural matrix as per the definition is determined for the Stephenson's chain and it is given by,

$$[S] = \begin{bmatrix} \begin{bmatrix} 0 & 1 & 0 & 0 & 1 & 0 \\ 1 & 0 & 0 & 0 & 0 & 1 \\ 0 & 0 & 0 & 0 & 1 & 1 \\ 0 & 0 & 0 & 0 & 1 & 1 \\ 1 & 0 & 1 & 1 & 0 & 1 \\ 0 & 1 & 1 & 1 & 0 & 0 \end{bmatrix} & \begin{bmatrix} 1 & 0 & 0 & 0 & 0 & 0 & 1 \\ 1 & 1 & 0 & 0 & 0 & 0 & 0 \\ 0 & 0 & 0 & 1 & 1 & 0 & 0 \\ 0 & 0 & 1 & 0 & 0 & 1 & 0 \\ 0 & 0 & 0 & 0 & 1 & 1 & 1 \\ 0 & 0 & 1 & 1 & 0 & 0 & 0 \end{bmatrix} \\ \begin{bmatrix} 1 & 1 & 0 & 0 & 0 & 0 \\ 0 & 1 & 0 & 0 & 0 & 0 \\ 0 & 0 & 0 & 1 & 0 & 1 \\ 0 & 0 & 1 & 0 & 0 & 1 \\ 0 & 0 & 1 & 0 & 1 & 0 \\ 0 & 0 & 0 & 1 & 1 & 0 \\ 1 & 0 & 0 & 0 & 1 & 0 \end{bmatrix} & \begin{bmatrix} 0 & 1 & 0 & 0 & 0 & 0 & 1 \\ 1 & 0 & 0 & 1 & 0 & 0 & 0 \\ 0 & 0 & 0 & 1 & 0 & 1 & 0 \\ 0 & 1 & 1 & 0 & 1 & 0 & 0 \\ 0 & 0 & 0 & 1 & 0 & 1 & 1 \\ 0 & 0 & 1 & 0 & 1 & 0 & 1 \\ 1 & 0 & 0 & 0 & 1 & 1 & 0 \end{bmatrix} \end{bmatrix}$$

Thus, the above structural matrix S forms the basis for calculating an index for addressing the problem of isomorphism in kinematic chains. The isomorphism index takes the form of (n+j+1) integers which are the characteristic coefficients of S. It may be noted that both incidence matrix and joint adjacency of a chain can be directly derived from its link adjacency matrix. The primary intention of creating the larger matrix such as S is to pack as much information as possible into the matrix representation of the chain from which the isomorphism index will be calculated.

It is interesting to note that one can think of the structural matrix of a given chain or graph as the vertex-adjacency matrix of yet another larger graph (for convenience, may be called structural graph) whose dimension is the sum of number of links and joints present in the given graph. Although a structural graph/chain is far from representing a kinematic chain, one can observe that there is a one-to-one relationship between the given graph and its corresponding structural graph. That is, one can uniquely lead to the other. Since the vertices and edges involved in the structural graphs are generally larger in number, the characteristics coefficients resulting from them are more varied and therefore contributes to the robustness of structural matrix based isomorphism test.

To illustrate the above consider the 4-bar chain in which the four vertices and edges are labeled as shown in Fig. 2(a).

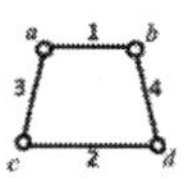

$$\begin{bmatrix} 0 & 0 & 1 & 1 \\ 0 & 0 & 1 & 1 \\ 1 & 1 & 0 & 0 \\ 1 & 1 & 0 & 0 \end{bmatrix} \quad \begin{bmatrix} 0 & 1 & 1 & 0 \\ 1 & 0 & 0 & 1 \\ 1 & 0 & 0 & 1 \\ 0 & 1 & 1 & 0 \end{bmatrix} \quad \begin{bmatrix} 1 & 1 & 0 & 0 \\ 0 & 0 & 1 & 1 \\ 1 & 0 & 1 & 0 \\ 0 & 1 & 0 & 1 \end{bmatrix} \quad \begin{bmatrix} 0 & 0 & 1 & 1 & 1 & 1 & 0 & 0 \\ 0 & 0 & 1 & 1 & 0 & 0 & 1 & 1 \\ 1 & 1 & 0 & 0 & 1 & 0 & 1 & 0 \\ 1 & 1 & 0 & 0 & 0 & 1 & 0 & 1 \\ 1 & 0 & 1 & 0 & 0 & 1 & 1 & 0 \\ 1 & 0 & 0 & 1 & 1 & 0 & 0 & 1 \\ 0 & 1 & 1 & 0 & 1 & 0 & 0 & 1 \\ 0 & 1 & 0 & 1 & 0 & 1 & 1 & 0 \end{bmatrix}$$

(a) b) [*LAM*] (c) [*JAM*] (d) [IM] (e) [*S*]

Fig. 2: Four-bar chain and its various matrix representations.

For the chain shown in Fig. 2a, Fig. 2b shows the link-adjacency matrix; Fig. 2c shows the joint adjacency matrix; Fig. 2d shows the link-joint incidence matrix; and finally, Fig. 2e shows the structural matrix that combines the three matrix representations.

Journal of Physics: Conference Series **2332** (2022) 012009 doi:10.1088/1742-6596/2332/1/012009

Treating the structural matrix as a chain or a linear graph, one can sketch the chain shown in Fig. 3a and the corresponding graph shown in Fig. 3b. It may be observed that the structural chain is a lot more intricate than the 4-bar chain from which the structural chain was derived. As noted in Mruthyunjaya [8], computation of characteristic coefficients of a given graph essentially involves summation of all possible paths present between any two vertices in the graph. As per the latter, it may be concluded that the more complex the graph becomes in terms of the number of vertices, edges and connectivity between vertices, there shall exist more number of paths between the vertices in the graph and the lengths of the paths also will be relatively long. This leads to higher order of rigor in the results obtained from such an isomorphism test.

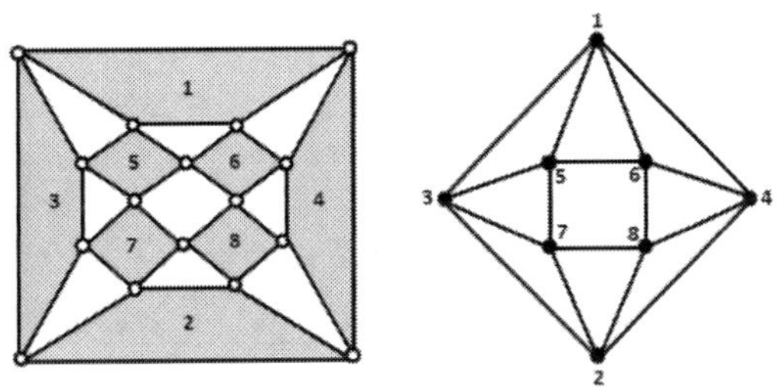

(a) (b)

Fig. 3: Sketches of the structural chain/graph as per structural matrix.

The characteristic coefficients of link-adjacency matrix for the 4-bar chain (see Fig.2b) is [1, 0,–4, 0, 0] while that of structural matrix (see Fig. 2e) is[1, 0,–16,–16, 52, 64,–48,–64, 0]. The relatively larger values of coefficients of the structural matrix primarily arise from the complexity of the corresponding graphs.

3. RESULTS AND CONCLUSION

Except for the categories of 10 link kinematic chains, namely the 10 link, 1 d.o.f. and 10 link, 3 d.o.f. cases (see also [8] and Table-1, Sl. No. 1 and 2), there have been no study in the literature to identify cospectral chains containing 11 or more links. This paper attempts to explore the existence of such chains.

The first step in this study is to enumerate chains containing up to 20 links and possessing up to 13 degrees of freedom [15]. Towards this end, the authors have used their own program for structural synthesis of SJKC structures. The isomorphism test used in [15] was based on the generation of an ordered set of N strings from an N link chain. Two chains were declared isomorphic if both yielded identical set of strings.

With the availability of complete collection of SJKCs, an attempt was made to compute and compare characteristic coefficients of the link-adjacency matrix of each chain. This exercise resulted in a collection of 797 cospectral chains in the categories of chains listed in Table–1 below. Due to lack of space the 797 cospectral chains obtained have not been sketched here. However, they can be made available to the interested upon request.

Table – 1: Occurrences of Cospectral Cases in SJKC Categories

Sl.No.	SJKC Category	Doublets	Triplets	# Cospectral cases
1	10 link, 1 d.o.f	2	0	4
2	10 link, 3 d.o.f	2	0	4
3	11 link, 2 d.o.f	17	1	37
4	11 link, 4 d.o.f	1	0	2
5	12 link, 1 d.o.f	183	14	408
6	12 link, 3 d.o.f	33	0	66
7	12 link, 5 d.o.f	2	0	4
8	13 link, 4 d.o.f	72	8	168
9	13 link, 6 d.o.f	2	0	4
10	14 link, 7 d.o.f	6	0	12
11	15 link, 8 d.o.f	2	0	4
12	16 link, 9 d.o.f	6	0	12
13	17 link, 10 d.o.f	6	0	12
14	18 link, 11 d.o.f	9	0	18
15	19 link, 12 d.o.f	8	0	16
16	20 link, 13 d.o.f	13	0	26

To check how the isomorphism tests based on degree matrix and structural matrix perform, characteristic coefficients of the degree and structural matrices of each of the above 797 cospectral chains were computed and compared. Except for the two pairs of 12 link, 1 d.o.f cospectral chains, isomorphism tests based on the degree and structural matrices proved to be satisfactory. The two cases of 12 link, 1 d.o.f chains that yielded identical coefficients are shown in figure 4. The corresponding coefficients are listed in Table–3.

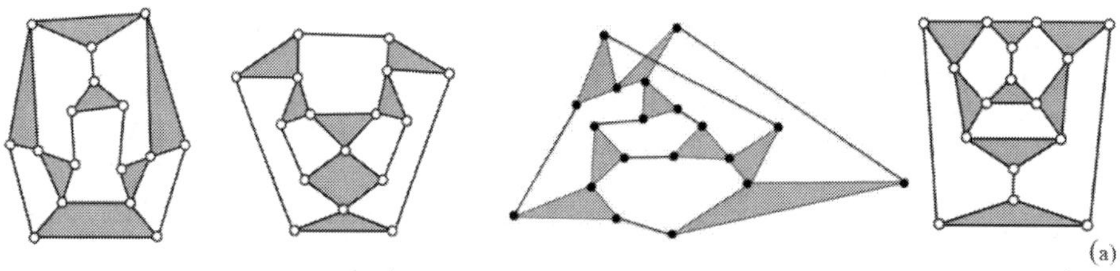

(a)

(b)

Fig. 4: 12-link 1 d.o.f cospectral chains

Table – 2: Characteristic Coefficients of Adjacency, Degree and Structural Matrix Representations

Fig – 2	Matrix Representation	Characteristic Coefficients
(a)	Link-adjacency matrix	1, 0, -16, 0, 87, 0, -206, 0, 207, 0, -64, 0, 4
	Degree Matrix	1, -12, -498, 2180, 63435, -134242, -3203882, 3166458, 63577824, -21285760, -307509248, 42827776, 320339968
	Structural Matrix	1, 0, -77, -110, 2376, 6498, -33949, -149946, 159265, 1654964, 1275972, -8327112, -17864605, 10024328, 70277213, 56152234, -81720200, -169089530, -52805519, 109632154, 109229383, 11126408, -32541276, -18176712, -2975008, 270400, 131904, 12160, 256
(b)	Link-adjacency matrix	1, 0, -16, 0, 90, -4, -226, 24, 252, -48, -96, 32, 0
	Degree Matrix	1, -12, -472, 2106, 60279, -144870, -3136176, 4873640, 68315328, -84376064, -465602560, 690636800, 0

	Structural Matrix	1, 0, -76, -104, 2342, 6176, -34103, -144986, 174687, 1652414, 1158115, -8709238, -18504469, 10567632, 76542245, 69818122, -77263956, -203033142, -122141592, 56432694, 109901501, 43167004, -8382632, -10622192, -2319168, 53312, 57472, 2304, -256

An observation that may be made from the pairs of cospectral chains shown in Fig. 2 is that they exhibit high level of symmetry. This may be the probable reason why even structural matrix based failed to detect structural dissimilarity in the chains.

In short, it may be observed from the work that, although the degree matrix size is (N x N), its performance is as good as structural matrix ((N+J) x (N+J)). Also by calculating and comparing the coefficients of joint adjacency matrix representation of the 408 cospectral chains in the 12-link 1 d.o.f category resulted in 10 failures only.

References

1. M.Huang, A.H.Soni, "Application to linear and non-linear graphs in structural synthesis of kinematic chain", ASME, Journal of Engineering for Industry Vol 95, 1973, pp. 525-532.
2. J.J.Uicker, A.Raicu, "Method for identification and recognition of equivalence of kinematic chains", Mechanism and Machine Theory Vol 10, 1975, pp. 375-383.
3. H.S. Yan, A.S. Hall, Linkage characteristic polynomials: definitions, coefficients by inspection, ASME J. Mech. Des. 103 (3) (1981) 578–584.
4. H.S. Yan, A.S. Hall, Linkage characteristic polynomials: assembly, theorems, uniqueness, ASME J. Mech. Des. 104 (1) (1982) 11–20.
5. T.S. Mruthyunjaya, A computerised methodology for structural synthesis of kinematic chains: Part 1 formulation, Mech. Mach. Theory 19 (6) (1984) 487–495.
6. T.S. Mruthyunjaya, A computerised methodology for structural synthesis of kinematic chains: Part 2 application to several fully or partial known cases, Mech. Mach. Theory 19 (6) (1984) 497–505.
7. T.S. Mruthyunjaya, A computerised methodology for structural synthesis of kinematic chains: Part 3 application to new case of 10- Link, three-freedom chains, Mech. Mach. Theory 19 (6) (1984) 507–530.
8. T.S. Mruthyunjaya, H.R. Balasubramanian, In quest of a reliable and efficient computational test for detection of isomorphism in kinematic chains, Mech. Mach. Theory 22 (2) (1987) 131–140.
9. W.J. Sohn, F. Freudenstein, An application of dual graphs to the automatic generation of the kinematic structures of mechanisms, ASME J. Mech. Des. 108 (1986) 92–398.
10. Vijayananda.K, "Computer Aided Structural Synthesis of Linkages and Epicyclic gear transmissions", Ph.D Thesis, IISc Bangalore 1994.
11. H.Ding, Z.Hang, "A new theory for the topological structure analysis of kinematic chains and its applications", Mechanism and Machine Theory Vol 42 (2007), pp. 1264-1279.
12. Manikandan.H, Vijayananda Kaup, Harish Babu, Identification of Cospectral Simple Jointed Kinematic Chains using Eigen Values and Eigen Vectors. (Paper Under Review).
13. S.C.Pilli, Vijayananda Kaup, S.F.Patil, "Structural Synthesis of Kinematic chains using Eigen values and Eigen vectors", 13thNational Conference on Mechanisms and Machines (NaCoMM07), NaCoMM-2007-073.
14. H. S. Yan and W. M. Hwang,J. Mech. Transmission. Automat. Design 105,658 (1983).

15. Manikandan H, Vijayananda Kaup, Harish babu, "A New Computational Methodology for Synthesis of Simple Jointed Kinematic chains", AIP Conf. Proc. 2134, 070003-1–070003-8.
16. Manikandan H, Vijayananda Kaup, Harish babu, "In quest of a reliable and efficient isomorphism index for simple jointed kinematic chains", AIP Conf. Proc. 2134, (2019) 070004-1-8.
17. Manikandan H, Vijayananda Kaup, Harish babu, "In quest of a reliable and efficient isomorphism index for multiple jointed kinematic chains", AIP Conf. Proc. 2134, (2019) 070005-1-11.
18. Manikandan H, Vijayananda Kaup, Harish babu, "A computational methodology for detection of distinct mechanisms from simple jointed kinematic chains", AIP Conf. Proc. 2341, (2021) 020026-1-7.
19. Manikandan H, Vijayananda Kaup, Harish babu, "A computational method for synthesis of Baranov truss from simple jointed kinematic chains", AIP Conf. Proc. 2341, (2021) 020027-1-6.
20. Manikandan H, Vijayananda Kaup, Harish babu, "Structural synthesis and analysis of 12-link, 3-freedom simple jointed kinematic chain", AIP Conf. Proc. 2341, (2021) 020031-1-14.

Rendering Abstract Notation of Kinematic Chains to Modified Franke's Notation Form

Manikandan H[1,4], Vijayananda Kaup[2], Harish Babu[3]

[1,4]Department of Mechanical Engineering (VTU-RC), CMR Institute of Technology-Bengaluru-560037, Karnataka, India
[2,3]Department of Mechanical Engineering, CMR Institute of Technology-Bengaluru-560037, Karnataka, India.

Abstract. To bring about innovation in the development of any product, it is required that the designer explores multitude design concepts right at the conceptual design phase of the product development cycle. At this phase, results of structural synthesis and analysis of mechanisms have been shown to serve as an invaluable tool in the attempt to infuse novelty in the design being conceptualized. There are several studies available in literature dealing with structural enumeration of kinematic chains. Invariably, all these studies culminate in reporting the results in an abstract form such as link adjacency matrix or its simpler variants like joints' lists, etc. Certainly, it will be more useful to designers if the chain were reported in the pictorial form since it will be more intuitive for designers to fruitfully relate what is in his mind as an unformulated idea to what is directly visible as a sketch. This paper presents the results of a computer based method that converts link adjacency matrix into modified Franke's notation format. The way to draw full-fledged chain from its modified Franke's notation form is explained. Chains containing 8 or more links and possessing 1 to 5 freedoms have been sketched and made available in the form of a PDF file.

Keywords: kinematic chains, link adjacency matrix, Frankie notation, degree of freedom

1. INTRODUCTION

It is a well-studied fact that the motion characteristics of any given linkage is almost entirely decided upon by the nature of interconnections between the constituent elements of the linkage. These elements take the form of kinematic links. A link is characterized by the number of points or locations (distinct or otherwise) through which it gets connected to other links of the same chain. In the conceptualization phase of machine design, however, the general motion pattern is assumed to be unaffected by the dimensional aspects of the mechanism. Specifically, the metric properties do not come in the way of the kinematic structure; degrees of freedom of the linkage; the type of freedom exhibited by the linkage; the distinct mechanisms derivable from the linkage; the distinct actuator(s) locations possible from a given mechanism; and so on.

The computerized structural synthesis methodologies and programs published thus far generally outputs the kinematic structures of linkages in the form of their algebraic representations. However, it will be desirable for mechanism designers to have a graphic depiction of the linkage structures so as to lend them a visual understanding of the link–joint interrelationship of the generated structures and an intuitive feel for their spatial relationships. This paper attempts to develop a program that outputs the sketches of the synthesized linkages in an intermediate format that closely resembles the actual kinematic chain.

2. LITERATURE REVIEW

Franke [1] was one of the few to introduce a concise graphical notation to represent KCs. His notation consisted of circles representing polygonal links, and lines representing strings of binary links that connect the polygonal links. The number i of an i-nary polygonal link is placed within the respective circle, and the integer indicating the number of consecutive binary links (including zero) in each string is placed next to the line representing that string. Fig.1 shows Franke's symbolic representation of an 8-link, 1-d.o.f KC.

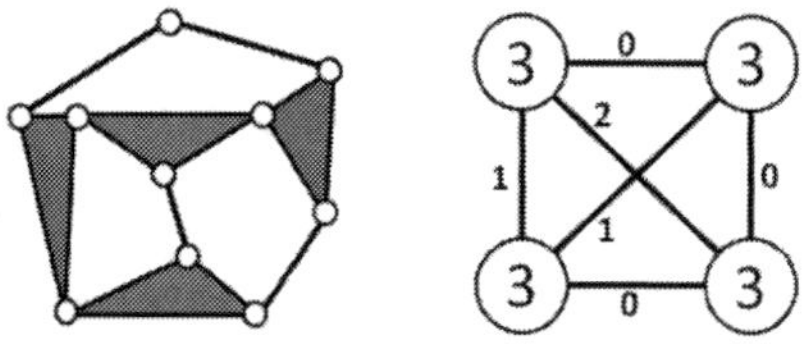

Fig. 1 8 link, 1 d.o.f Kinematic Chain and its Franke's Notation

Journal of Physics: Conference Series **2332** (2022) 012010 doi:10.1088/1742-6596/2332/1/012010

Several researchers have employed Franke's notation to determine the distinct permutations of certain classes of mechanisms (Davies and Crossley [2], Haas and Crossley [3], Soni [4]). Davies and Crossley [2] used Franke's notation to detect structural isomorphism of the KCs although it can be drawn in different ways. In his paper Mruthyunjaya [5] observes that structural synthesis programs developed generally present the chains by their algebraic representations. He mentions that it is desirable to obtain graphic display of chains directly by means of a computer program. This is to help the designer with a visual understanding and intuitive feel of the interrelationship between the links and joints of the enumerated structures. Dobrjanskyj and Freudenstein [6] were the first to attempt the automatic sketching of the kinematic chains but their sketching algorithms without avoiding of link crossings. Woo [7] presented a rule based approach for line graphs without edge crossings that emphasized the importance of determining independent loops of the graph. Olson et al. [8] presented a procedure for transforming the conventional graph of chain into its line graph and an algorithm to sketch the line graph with crossing edges removed from the complete sub-graphs and the complete sub-graphs replaced by hatched triangles and polygons.
Cheng and Hoeltzel [9] employ a combinatorial approach for algorithmically solving the numerically continuous chain sketching problem as a discrete domain problem.
Belfiore and Pennestri [10] present a method for fully automatic sketching of a simple-jointed chain given its link adjacency list. Their method is said to have produced the automatic sketching of the single-freedom chains with 8 and 10 links without any occurrence of crossing links. Mauskar and Krishnamurthy [11] present a method similar to the one in [9] but simpler, more systematic, fully automated and believed to be computationally more efficient.

3. NOTATION, CONVENTIONS AND CONVERSION TO KC

Fig. 2 shows a 11-link, 4-freedom simple jointed kinematic chain with separation link. In addition, the distinct mechanisms from the chain and the link-assortment are also reported.

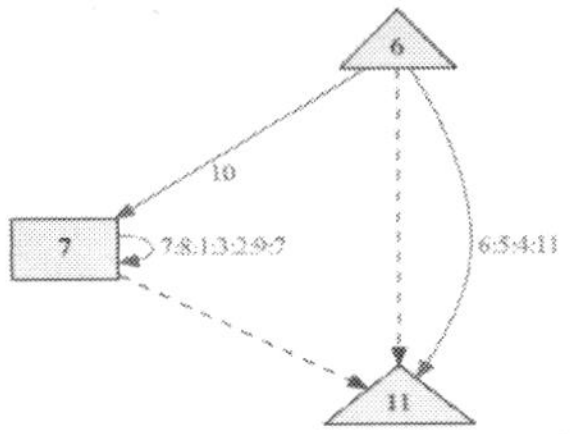

Kinematic chain 1

Planar [[1, 2], [3], [4], [5], [6], [7], [8, 9], [10], [11]]

Fig. 2 11-link, 4-freedom Simple Jointed Kinematic Chain with Separation Link

Journal of Physics: Conference Series **2332** (2022) 012010 doi:10.1088/1742-6596/2332/1/012010

The chain notation comprises graphic and text parts. Details of the graphic part is as follows:

1. Polygons having 3, 4, 5, 6,.... sides representing ternary link, quaternary link, pentagonal link, hexagonal link, etc. respectively. For example, triangles labelled 6 and 11 in Fig.1 represent ternary links; a quadrilateral labelled 7 represents a quaternary link.
2. Any line connecting two polygon links *a* and *b* represent one or more binary links between the links *a* and *b*. For example, the line labelled 10 in Fig.1 interposed between triangle 6 and quadrilateral 7 represents binary link 10 between ternary link 6 and quaternary link 7. Likewise, the line carrying the label 6:5:4:11 in Fig.1, interposed between ternary links 6 and 11 represents a dyad 5, 4 interposed between the ternaries 6 and 11.
3. Any line leaving and returning to the same polygon represents a chain with fractionated freedom. The latter can be due to the presence of a separation link or a separation pair. For example, the line carrying the label 7:8:1:3:2:9:7 in Fig.1, emerging and culminating in quaternary link 7 represents series of binary links 8, 1, 3, 2, 9 in the listed order.
4. The broken line between the ternaries 6 and 11 indicates direct connection between them. Likewise, quaternary link 7 and ternary link 11 are directly connected.

Interpretation of text part is as follows:

1. Line-1 denotes the sequence number of the enumerated chain.
2. The second line indicates whether the chain is planar/non-planar, and the links that yield distinct mechanisms. For the example in Fig.1, line 2 tells that the chain is planar, and the mechanism derived by fixing links 1 and 2 (enclosed within the same parenthesis) are identical; likewise, mechanism derived by fixing links 8 and 9 are identical. Thus, there can be 09 distinct mechanisms possible by fixing links 1, 3, 4, 5, 6, 7, 8, 10, 11.
3. Line-3 provides details of link assortment in the chain. 4S*1 means that there is only 1 four-sided (4S) polygonal link or the quaternary. 3S*2 tells there are two ternaries in the chain. Likewise, there 2S*8 means the presence of 8 binary links.

4. CONVERSION TO KINEMATIC CHAINS

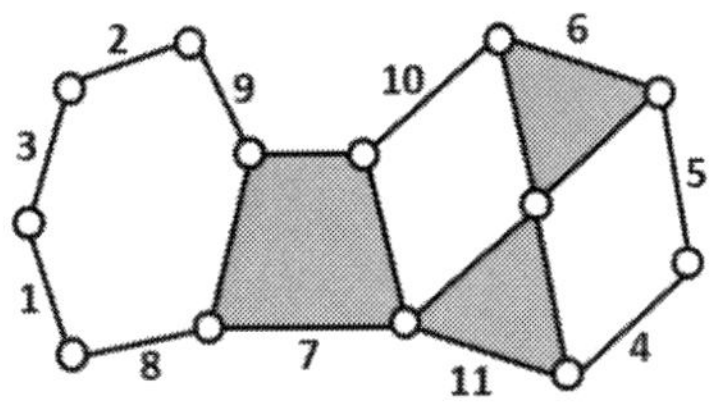

Fig. 3 Notation in Fig.2 Converted to Simple Jointed Kinematic Chain

Using the conventions explained above, the KC notation presented in Fig.1 is converted to a SJKC as shown in Fig.3. From visual inspection, it may be noted that links 8 and 9 are identically disposed. So is with links 1 and 2. Hence, the nine links that yield distinct mechanisms are {1, 3, 4, 5, 6, 7, 8, 10, 11}.

The following example shown in Fig.4 illustrate a KC containing a separation pair between links 7 and 9. The corresponding KC is shown in Fig.4. Again, from visual inspection it may be noted that links 1, 2; links 4, 5; links 10, 11 have identical dispositions. This chain can yield 8 distinct mechanisms by fixing links {1, 3, 4, 6, 7, 8, 9, 10}

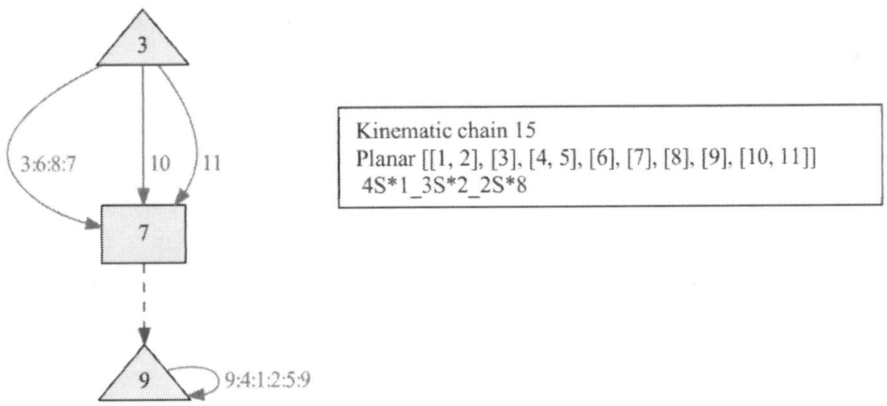

Fig. 4 11-link, 4-freedom Simple Jointed Kinematic Chain with Separation Pair

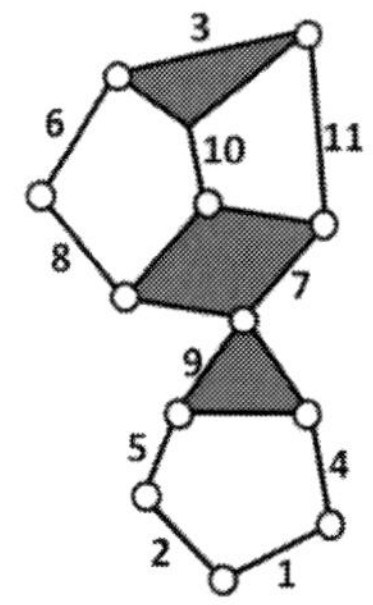

Fig 5 Notation in Fig.4 Converted to Simple Jointed Kinematic Chain

5. RESULTS AND DISCUSSION

In this research work we developed a Python program that synthesizes simple jointed kinematic chains belonging to categories of chains containing 6 to 20 links and possessing up to 15 freedoms. The output of the program is a PDF document containing details of the enumerated chain as shown in Fig.2 and Fig.4. The method adopted to synthesize KCs involves systematic addition of predetermined link-patterns to "parent KCs" belonging to the categories consisting of fewer number of links so as to generate chains possessing one loop more than its predecessor (parent KC) [].

The program was used to enumerate the following categories of KCs. Output of the program is a PDF file available with the authors which can be shared on request. The following table shows the number of kinematic chains synthesized for authors [12,13].

Table 1 Simple Jointed kinematic chains and mechanisms

Sl. No.	Kinematic Chain Category		No. of KCs	No. of Mechanisms
	No of Links	Degree of Freedom		
1	6	1	2	5
2	7	2	4	14
3	8	1	16	71
4	8	3	7	26

5	9	2	40	254
6	9	4	10	47
7	10	1	230	1834
8	10	3	98	684
9	10	5	14	72
10	11	2	839	7898
11	11	4	189	1564
12	11	6	19	111
13	12	1	6856	75397
14	12	3	2422	25125
15	12	5	354	3199
16	12	7	24	154

6. CONCLUSION

A pictorial representation of the chain will be more useful to designers engaged in mechanism and machine design. Rather than presenting kinematic chains in an abstract form such as a matrix or joint details, this work has developed a Python program that converts the matrix into a modified form of Franke's notation. The latter can then be easily converted to actual kinematic structure. Along with the kinematic structure, the program indicates distinct mechanisms possible from the chain and the assortment of links in the chain. The program has been created in such a way that all the synthesized chains are written into a PDF file for easy exchange. The collection of chains can be made available to interested researchers and designers by communicating with the authors of this paper or to the journal editors.

References

1. Franke, R. "*Von Aufbau der Getriebe*", Volume I, Benth - Vertrich, Oxford, 1943.
2. Davies, T. H. and Crossley, F. R. E. "Structural Analysis of Plane Linkages by Franke's Condensed Notation," *Journal of Mechanisms*, **1**:171-183, 1966.
3. Haas, S. L. and Crossley, F. R. E. "Structural Synthesis of a Four-Bit Binary Adding Mechanism," *ASME Journal of Engineering for Industry*, Series B, **91**: 240-250, 1969.
4. Soni, A. H. "Structural Analysis of Two General Constraint Kinematic Chains and

their Practical Application," *ASME Journal of Engineering for Industry*, Series B, **93**: 231-238, 1971.

5. Mruthyunjaya, T. S. "Kinematic structure of mechanisms revisited," *Mechanism and Machine Theory*, **38,** 279–320, 2003.
6. L. Dobrjanskyj, F. Freudenstein, Some applications of graph theory to structural analysis of mechanisms, J. Eng. Indust., ASME Trans., Series B 89 (1967) 153–158.
7. L.S. Woo, An algorithm for straight-line representation of simpler planar graphs, J. Franklin Inst. (1969) 197–208.
8. D.G. Olson, T.R. Thompson, D.R. Riely, A.G. Erdman, Algorithm for automatic sketching of planar kinematic chains, J. Mechanisms, Transmissions, Automat. Des. 107 (1985) 106–111.
9. W.H. Cheng, D.A. Hoeltzel, A combinatorial approach for the automatic sketching of planar kinematic chains and epicyclic gear trains, J. Mech. Des., ASME Trans. 112 (1990) 6–15.
10. N.P. Belfiore, E. Pennestri, Automatic sketching of planar kinematic chains, Mech. Mach. Theory 29 (1994) 177–193.
11. S. Mauskar, S. Krishnamurthy, A loop configuration approach to automatic sketching of mechanisms, Mech. Mach. Theory 31 (1996) 423–437.
12. Manikandan.H, Vijayananda Kaup, Harishbabu, A computational methodology for detection of distinct mechanisms from simple jointed kinematic chains, AIP Conference Proceedings 2341, 020026 (2021).
13. Manikandan.H, Vijayananda Kaup, Harishbabu, A new computational methodology for the synthesis of simple jointed kinematic chains, AIP Conference Proceedings 2134, 070003 (2019).

Influence of Body Acceleration and Slip Velocity on Fluid Flow in a Multi-Stenotic Artery

Asha K N[1], Neetu Srivastava[2]

[1]Department of Mathematics, Research Scholar (Amrita Vishwa Vidyapeetham University), CMR Institute of Technology, Bengaluru, Karnataka.

[2]Associate Professor, Department of Mathematics, Amrita School of Engineering, Bengaluru, Amrita Vishwa Vidyapeetham University, India.

asha.kn@cmrit.ac.in

Abstract. Blockage of arteries by the formation of plaques causing stenosis is a major health problem among human beings.In this study, a model is developed for the flow through a multi-stenosed elastic artery segment under the effect of body acceleration and slip velocity. Fluid is regarded as Non-Newtonian Bingham Plastic Fluid. The equations representing the model are solved analytically using Perturbation Method. Result is discussed for velocity, wall shear stress and volumetric flow rate by varying several parameters like pulse frequency, slip parameter and Womersley Frequency parameter graphically.

1. Introduction

Coronary arteries carry oxygenated blood. The blockage of these arteries by the formation of plaque decreases the flow of blood. This plaque may also burst and cause blood clot. This may trigger a heart attack or stroke. Hence, formation of arteries is considered as one of the major health issues these days as the changed life style and food habits are causing the increased formation of plaque in human arteries. Because of the increased need, blood flow in a blocked artery has become one of the major topics of research since last few decades.

Extensive research has been carried out on overlapped stenotic artery. Asha and Srivastava Neetu [5] studied and documented the work of various researchers in this field. The blood circulation in an artery with overlapped stenosis is studied by Prasad et al. [1]. Results conclude that resistance, pressure drop and wall shear stress increase with the stenosis height.Chakravarty and Mandal [2] examined the arterial movement of blood in the case of overlapped stenosis. It is observed that resistance to flow is largest when the stenosis height is highest. Ismail et al. [3] also considered overlapped stenosis during the motion of blood treating blood as non-Newtonian model. It is noticed that the resistance and impedance are larger compared to Newtonian model. Riahi et al. [4] during their study found that the impedance and pressure gradient are more in the case of mild stenosis. Several authors [7, 8, 9] have studied the presence of slip velocity and have discussed its importance.

In this study, the flow of Bingham Plastic fluid model in a multi-stenotic artery is considered under slip velocity and body acceleration. The body fluid is considered to be non-Newtonian and the artery is elastic in nature. Bingham plastic is a material that behaves as a viscous fluid at high stress, like tooth paste and chocolate.

2. Problem Formulation

Consider a blood flow past segment of a flexible artery which involves the presence of external periodic body acceleration. Let the flow be considered as one dimensional pulsatile of a Casson fluid. Stenosis developed in this segment is considered to be radially symmetric. The geometry of the segment is as shown in Fig 1.

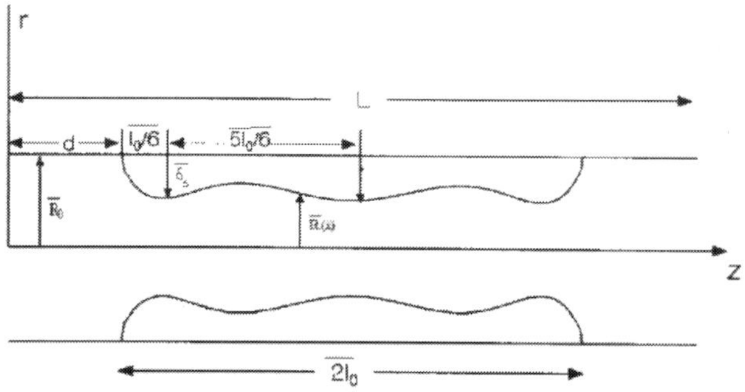

Figure 1. Configuration of a multi-stenotic arterial segment.

Geometry of the stenosis, from [6], is given by

$$\frac{\bar{R}(\bar{z})}{\bar{R_0}} = \left[1 - \frac{\bar{\delta_s}}{5005a\bar{l}_0^6}\left\{\frac{668662}{9}(\bar{z}-\bar{d})\bar{l}_0^5 - 370281(\bar{z}-\bar{d})^2\bar{l}_0^4 + 743344(\bar{z}-\bar{d})^3\bar{l}_0^3 - 698476(\bar{z}-\bar{d})^4\bar{l}_0^2 + 307584(\bar{z}-\bar{d})^5\bar{l}_0 - 51264(\bar{z}-\bar{d})^6\right\}\right]a_1(t), \qquad \bar{d} \le \bar{z} \le \bar{d} + 2\bar{l}_0 \quad (1)$$

$$= a_1(t), \qquad otherwise$$

Here $\bar{\delta_s}$ is the critical size of the stenosis, $\bar{R}(\bar{z})$and $\bar{R_0}$ are the radii of the artery in the critical region and non-stenotic region, $2\bar{l}_0$ is the length of the stenosis, divided as $z = d + (\bar{l}_0/6), d + \bar{l}_0$ *and* $d + (11/6)\bar{l}_0$ and
$a_1(t) = 1 - (cos\ \bar{\omega}\bar{t} - 1)\beta e^{\beta\bar{\omega}\bar{t}}$. Here,$\bar{\omega}$ represents the angular frequency, β is a constant parameter and t is a generic parameter.

The body acceleration $\bar{F}(\bar{t})$along the axial direction is

$$\bar{F}(\bar{t}) = a_0 \cos(\bar{\omega_b}\bar{t} + \varphi) \quad (2)$$

where a_0 is the amplitude, $\bar{\omega_b} = 2\pi\bar{f_b}$; $\bar{f_b}$is the frequency of body acceleration. This frequency $\bar{f_b}$ is assumed to be small.

Pressure gradient is given by

$$-\frac{\partial\bar{p}}{\partial\bar{z}} = A_0 + A_1 cos(\bar{\omega_p}\bar{t}), t \ge 0 \quad (3)$$

$\bar{\omega_p} = 2\pi\bar{f_p}$; $\bar{f_p}$ is the pulse rate frequency.

Equations representing the current fluid flow model, as given in [7], are given by

$$\bar{\rho}\left(\frac{\partial\bar{u}}{\partial\bar{t}}\right) = -\left(\frac{\partial\bar{p}}{\partial\bar{z}}\right) - \left(\frac{1}{r}\right)\frac{\partial}{\partial\bar{r}}(\bar{r}\bar{\tau}) + \bar{F}(\bar{t}) \quad (4)$$

$$\frac{\partial \bar{p}}{\partial \bar{r}} = 0 \tag{5}$$

where $\bar{u}$ is the velocity along the axial direction, $\bar{p}, \bar{\rho}, \bar{t}, \bar{\tau}$ and $\bar{F}(\bar{t})$ are pressure, density, time, shear stress and body acceleration respectively.

Equations representing the Bingham Plastic Fluid are:

$$\bar{\tau} = \overline{\tau_y} - \bar{\mu}\left(\frac{\partial \bar{u}}{\partial \bar{r}}\right); \ if\ \bar{\tau} \geq \overline{\tau_y}$$

$$\frac{\partial \bar{u}}{\partial \bar{r}} = 0; \quad if\ \bar{\tau} < \overline{\tau_y} \tag{6}$$

Boundary conditions are given by:

$$\bar{u} = \overline{u_s}\ at\ \bar{r} = \bar{R}(\bar{z}) \tag{7}$$

$$\bar{\tau} = 0\ at\ \bar{r} = 0 \tag{8}$$

Following non-dimensional terms will be introduced.

$$u = \frac{\bar{u}}{A_0\bar{R}_0^2/4\bar{\mu}}\ ,\ z = \frac{\bar{z}}{\bar{R}_0}\ ,\ R(z) = \frac{\bar{R}(\bar{z})}{\bar{R}_0}\ ,\ r = \frac{\bar{r}}{\bar{R}_0}\ ,\ t = \bar{t}\overline{w_p}\ ,\ \omega = \frac{\overline{\omega_b}}{\overline{\omega_p}}\ ,\ \delta_s = \frac{\overline{\delta_s}}{\bar{R}_0}$$
$$,\ u_s = \frac{\overline{u_s}}{A_0\bar{R}_0^2/4\bar{\mu}}\ ,\ \tau = \frac{\bar{\tau}}{A_0 R_{0/2}}\ ,\ \alpha^2 = \frac{\bar{R}_0^2\overline{w_p}\bar{\rho}}{\bar{\mu}}\ ,\ e = \frac{A_1}{A_0}\ ,\ B = \frac{a_0}{A_0}\ ,\ \theta = \frac{\overline{\tau_y}}{A_0 R_{0/2}} \tag{9}$$

After non-dimensionalizing, equation (4) becomes,

$$\alpha^2\left(\frac{\partial u}{\partial t}\right) = 4(1 + ecost) + 4Bcos(wt + \varphi) - \left(\frac{2}{r}\right)\frac{\partial}{\partial r}(r\tau) \tag{10}$$

Here, $\alpha^2 = \frac{\omega_p R_0^2}{(\mu/\rho)}$

Equation (6), now becomes,

$$\tau = \theta - \frac{1}{2}\frac{\partial u}{\partial r} \quad ; \quad \tau \geq \theta$$

$$\frac{\partial u}{\partial r} = 0 \quad ; \quad \tau < \theta \tag{11}$$

Boundary conditions reduces to

$$u = u_s\ when\ r = R(z) \tag{12}$$

$$\tau \text{ is finite when} r = 0 \tag{13}$$

The geometry of the stenosis, after non-dimensionalising, is given by

$$R(z) = \left[1 - \frac{\delta_s}{5005al_0^6}\left\{\frac{668662}{9}(z-d)l_0^5 - 370281(z-d)^2 l_0^4 + 743344(z-d)^3 l_0^3 - 698476(z-d)^4 l_0^2 + 307584(z-d)^5 l_0 - 51264(z-d)^6\right\}\right] a_1(t), \quad for\ \ d \leq z \leq d + 2l_0$$
$$= a_1(t), \quad otherwise. \tag{14}$$

Non-dimensional volumetric flow rate is

$$Q = 4\int_0^{R(z)} ru\, dr \tag{15}$$

3. Solution

We solve this problem by perturbation method.

Let velocity and wall shear stress expressions be given by

$$u = u_0 + \alpha^2 u_1 + \cdots \tag{16}$$
$$\tau = \tau_0 + \alpha^2 \tau_1 + \cdots \tag{17}$$

Substituting the equations (16) and (17) in equation (10), solving and then applying the conditions (12) and (13), we obtain,

$$u_0 = u_s + f(t)(R^2 - r^2) - 2k^2 f(t)(R - r) \tag{18}$$

$$\tau_1 = -\frac{f'(t)}{24}[6R^2r - 3r^3 - 4k^2(3Rr - 2r^2)] \tag{19}$$

$$u_1 = \frac{f'(t)}{12}\left[3R^2r^2 - \frac{3}{4}(r^4 + 3R^4) + 2k^2\left(\frac{5}{3}R^3 - 3Rr^2 + \frac{4}{3}r^3\right)\right] \tag{20}$$

where $f(t) = [(1 + e\cos\ t) + B\cos(\omega t + \varphi)]$ and $k^2 = \frac{\theta}{f(t)}$

Using (18) and (20) in equation (16), we get the expression for velocity as

$$u = u_s + f(t)(R^2 - r^2) - 2k^2 f(t)(R - r) + \alpha^2\frac{f'(t)}{12}\left[3R^2r^2 - \frac{3}{4}(r^4 + 3R^4) + 2k^2\left(\frac{5}{3}R^3 - 3Rr^2 + \frac{4}{3}r^3\right)\right] \tag{21}$$

Substituting the obtained results in equation (17),

Wall shear stress, $\tau_\omega = f(t)R - \alpha^2\frac{f'(t)R^2}{24}(3R - 4k^2)$ (22)

Volumetric flow rate is given by

$$Q = 2u_sR^2 + f(t)R^4 - \frac{4}{3}k^2 f(t)R^3 + \alpha^2\frac{f'(t)}{12}\left(\frac{14}{5}k^2R^5 - 2R^6\right) \tag{23}$$

4. Results and Discussion

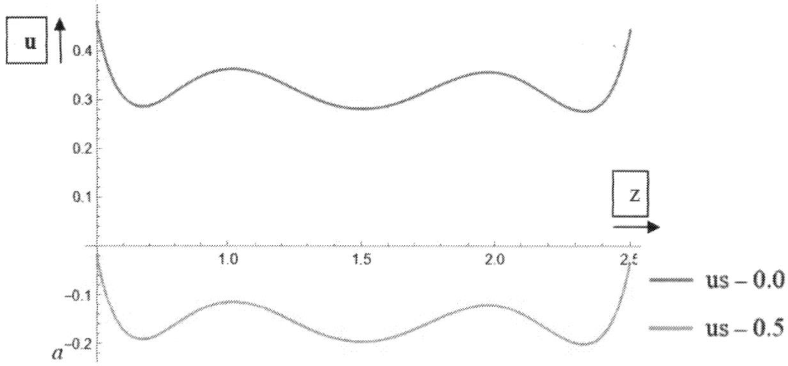

Figure 2. Velocity along axial axis for u_s= 0 and 0.5

The change in velocity along the axial axis for the value of slip velocity u_s= 0 and 0.5 in the stenotic region is shown in Fig 2. The velocity attains its minimum in the region when the stenosis height is largest. As there is a decrease in the stenosis height, there is an increase in the velocity. We can also notice that the velocity decreases as the slip velocity increases.

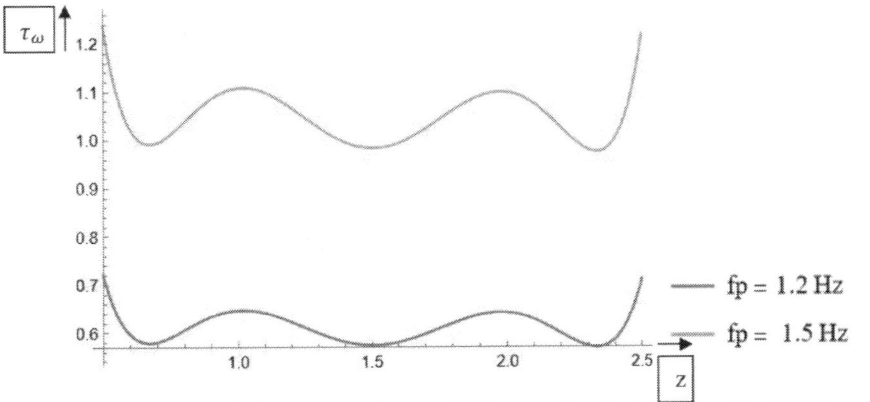

Figure 3. Wall shear stress along axial direction for fp=1.2 Hz and fp=1.5 Hz

The change in wall shear stress along the axial axis for two different frequencies is represented graphically in Fig 3. This graph depicts that the wall shear stress is least when the stenosis height is maximum. It is also observed that the wall shear stress increases as the value of pulse frequency increases.

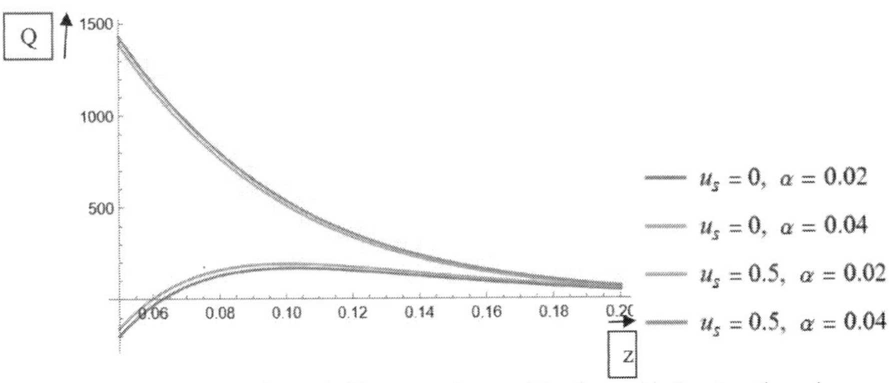

Figure 4. Volumetric Flow rate along axial axis outside the stenotic region

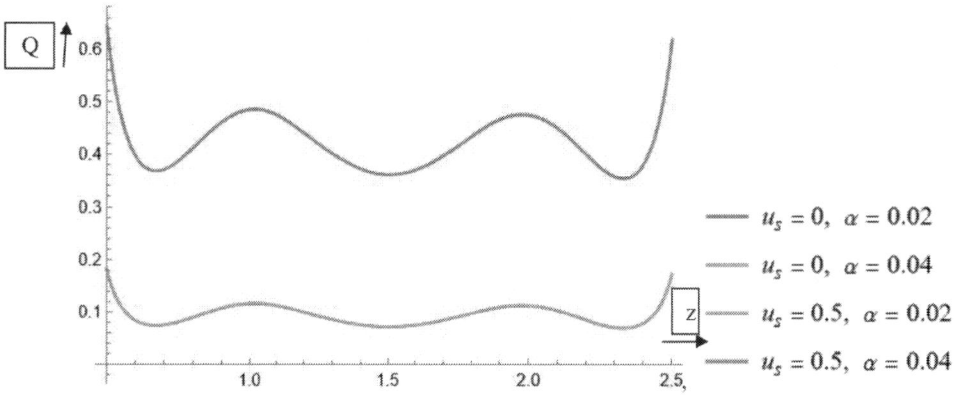

Figure 5. Volumetric Flow rate along axial axis in the stenotic region.

Fig 4 and Fig 5 shows the variation of volumetric flow rate inside and outside the stenotic region for different values of slip velocity and Womersley frequency parameter. The flow rate decreases in the stenotic region with the onset of stenosis and attains its minimum when the stenosis height is maximum. Flow rate is maximum for $u_s = 0.5$ than that of $u_s = 0$.

5. Conclusion

In this study, we have discussed the pulsatile flow through a multi-stenosed artery under slip velocity and body acceleration. Blood is treated as a non-Newtonian fluid by considering Bingham-Plastic fluid Model. Analytical solution is obtained for the flow parameters like velocity, wall shear stress and volumetric flow. It is observed that axial velocity decreases and the volumetric flow rate increases as there is an increase in slip velocity. This infers the importance of slip velocity in this model. It is also noticed that the pulse frequency enhances the wall shear stress. The current study may be extended by considering the presence of nano particles considered for the treatment of stenosis. This model helps the medical practitioners and the researchers in the field by providing relevant results.

A Nomenclature

Symbol	Meaning
$\overline{u_s}$	Slip velocity at the wall of the stenosis
$\overline{\tau_y}$	Yield stress
$\bar{\mu}$	Viscosity of blood
A_0	Steady state pressure gradient
A_1	Amplitude of the fluctuating component
$\overline{f_p}$	Pulse rate frequency
$\overline{f_b}$	Frequency of body acceleration
α	Womersley frequency parameter

References

[1] K. Maruthi Prasad, T. Sudha, M. V. Phanikumari, Investigation of Blood Flow through an Artery in the Presence of Overlapping Stenosis, Journal of Naval Architecture and Marine Engineering June, 2017

[2] S. Chakravarty and P. K. Mandal, Mathematical Modelling of Blood Flow Through an Overlapping Arterial Stenosis, Ma&l. Comput. Modelling Vol. 19, No. 1, pp. 59-7'0, 1994

[3] Zuhaila Ismail, Ilyani Abdullah, Norzieha Mustapha, Norsarahaida Amin, A power-law model of blood flow through a tapered overlapping stenosed artery, Applied Mathematics and Computation 195 (2008) 669–680

[4] Daniel N. Riahi, Ranadhir Roy, Sam Cavazos, On arterial blood flow in the presence of an overlapping stenosis, Mathematical and Computer Modelling 54 (2011) 2999–3006

[5] Asha K N and Neetu Srivastava, Geometry of stenosis and its effects on the blood flow through an artery - A theoretical study, AIP Conference Proceedings 2375, 030006 (2021)

[6] S. Chakravarty and A. KR. Sannigrahi, A Nonlinear Mathematical Model of Blood Flow in a Constricted Artery Experiencing Body Acceleration, Mathematical and Computer Modelling 29 (1999) 9-25

[7] S. U. Siddiqui, Geeta, Sapna, Effect of Body Acceleration and Slip Velocity on the Pulsatile Flow of Casson Fluid through Stenosed artery, e-Journal of Science & Technology (e-JST)

[8] R.Ponalagusamy, Blood flow through an artery with mild stenosis: a two layered model,different shapes of stenosis and slip velocity at the wall, J.Appl.Sci.7(7)(2007)1071–1077.

[9] Biswas,U.S.Chakraborty,Two layered pulsatile blood flow in a stenosed artery with body acceleration and slip at wall, AAM:Int.J.5(2)(2010)303–320.

Analysis of Copper nanoparticle suspension in a stenosed non-permeable vasculature in presence of magnetic field

Asha K N[1], Neetu Srivastava[2]

[1]Research Scholar, Department of Mathematics, Amrita School of Engineering, Bengaluru, Amrita Vishwa Vidyapeetham, India, 560035

[2]Associate Professor, Department of Mathematics, Amrita School of Engineering, Bengaluru, Amrita Vishwa Vidyapeetham, India, 560035

asha.kn@cmrit.ac.in

Abstract. In the present article, impact of shape of Copper nanoparticle with water as base fluid on the fluid flow criteria in the congested non-permeable arteries is analyzed under the magnetic effect. Mathematically, a significant response is recorded when various shapes of nanoparticles are considered in the stenosed vasculature. To explore the effect of different shaped nanoparticles, extensive range of nanofluid volume fractions, magnetic parameters and shape factors of nanoparticle is utilized. The significant parameters are solved using analytical method as well as analyzed using the graphs. It is found that the laminar nanoparticles show finer results for flow properties like flow rate in the stenosed artery and the impedance resistance.

1. Introduction

Vascular dynamics relates the physical action by humans to the blood flow. Cardiovascular complication and its treatment due to formation of plaques in the human arteries is medically focused area due to its rapid asymptotic growth. To deal with this, nanoparticles have proved their therapeutic properties. The nanoparticle technology has shown promising results by providing an improved and cheaper treatment with fewer side effects. Along with their chemical compositions, they play a significant role in effecting the blood flow in diseased artery because of their physical, chemical and thermal properties. Metallic nanoparticles have higher thermal conductivity as compared to regular fluids such as water, oil etc. and also have significant role of shape factors in affecting the flow. Due to these properties, it was considered to treat the stenosed area in the diseased artery.

Researchers [1-9] have analyzed critically the mathematical modeling for the blood flow characteristics in the stenosed arteries assuming Newtonian and non-Newtonian nature of blood. They developed the significant medical results. These results have also demonstrated the effect of stenosis over the crucial flow parameters. Taking into an account the presence of plaques in arteries, some researchers [3-7] have analyzed the tapered arteries with non-Newtonian behavior of blood and successfully depicted the flow characteristics through significant parameters.

Extensive research is done to study the effect of injecting nanoparticles into a stenosed artery. Akbar et al. [8] studied flow of nano fluid in a permeable artery with stenosis. The effect of Cu-Nanoparticles under magnetic field through a blocked artery is studied by Akbar and Mustafa [10].

Akbar and Butt [11] analyzed the effect of magnetic flux on flow of blood in an arterial segment containing stenosis with injected Cu-nanoparticles. Ahmed and Nadeem [12] examined the impact of shapes of the nano-particles in a stenosed artery containing a catheter. The effect of different shaped nanoparticles on the movement of blood is investigated by Asha and Srivastava Neetu [13]. [14]-[18] have studied the similar cases and have contributed to the results related to using nanoparticles and magnetic field for the treatment of diseased arteries.

Available literature presents the study of nanoparticles in various types of vasculatures. Injecting nanoparticles and analyzing the developed flow pattern with various type of conditions in permeable artery is done extensively. According to authors' knowledge, the behavior of shape of the copper nanoparticles under the magnetic effect through stenosed impermeable artery (Fig1) is not inspected so far. In this study, flow of blood through impermeable stenosed arteries using Cu-water nanofluid is being considered.

2. Formulation of Problem

Consider the movement of blood in a segment of an artery having a stenosis that is symmetric radially and axially non-symmetric. The configuration of the stenosis in the arterial segment is as shown in Figure 2 and is mathematically given by

$$\frac{R(z)}{R_0} = \begin{cases} 1 - \xi\left[L_0^{p-1}(z-d) - (z-d)^p\right], & d \le z \le d + L_0 \\ 1, & otherwise \end{cases} \quad (1)$$

Here, R(z) and R_0 are the radii of the artery having stenosis and not having stenosis respectively. $p \ge 2$ is shape parameter, L_0 and d are the length and location of the segment. ξ is a parameter that is given by

$$\xi = \frac{\delta}{R_0 L^p} \frac{p^{p/(p-1)}}{(p-1)}$$

where $\delta$is the greatest height of the stenosis at $z = \frac{(d+L_0)}{p^{1/(p-1)}}$

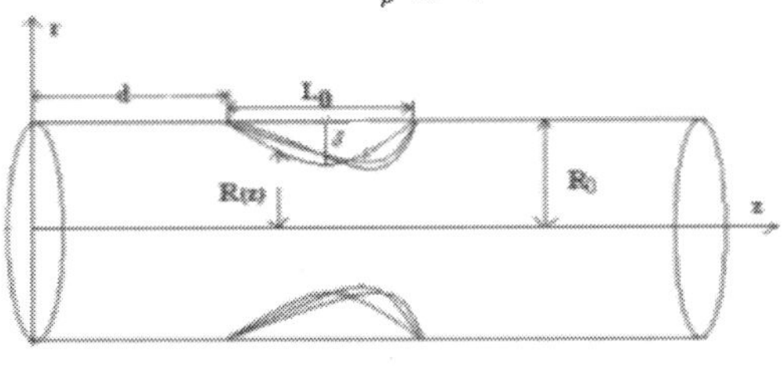

Figure 1. Geometry of radially symmetrical artery

The governing equations for incompressible nano fluid are

$$\frac{1}{r}\frac{\partial(rv)}{\partial r} + \frac{\partial u}{\partial z} = 0 \quad (2)$$

$$\rho_{nf}\left(v\frac{\partial v}{\partial r} + u\frac{\partial v}{\partial z}\right) = -\frac{\partial p}{\partial r} + \mu_{nf}\frac{\partial}{\partial r}\left(2\frac{\partial v}{\partial r}\right) + \mu_{nf}\frac{\partial}{\partial z}\left(2\frac{\partial v}{\partial z} + \frac{\partial u}{\partial r}\right) \quad (3)$$

$$\rho_{nf}\left(v\frac{\partial u}{\partial r} + u\frac{\partial u}{\partial z}\right) = -\frac{\partial p}{\partial z} + \mu_{nf}\frac{\partial}{\partial z}\left(2\frac{\partial u}{\partial z}\right) + \frac{\mu_{nf}}{r}\frac{\partial}{\partial r}\left[r\left(\frac{\partial v}{\partial z} + \frac{\partial u}{\partial r}\right)\right] + \rho_{nf}g\alpha(\bar{T} - T_0) - \sigma B_0^2 u \quad (4)$$

$$\left(\rho_{cp}\right)_{nf}\left(v\frac{\partial T}{\partial r} + u\frac{\partial T}{\partial z}\right) = k_{nf}\left[\frac{\partial^2 T}{\partial r^2} + \frac{1}{r}\frac{\partial T}{\partial r} + \frac{\partial^2 T}{\partial z^2}\right] + Q_0 \quad (5)$$

subject to the conditions depicting the impermeable flow through the artery,

$$u = 0 \; at \; r = \frac{R(z)}{R_0} \tag{6a}$$

$$\frac{\partial u}{\partial r} = 0, \qquad \frac{\partial T}{\partial r} = 0 \; at \; r = 0, \tag{6b}$$

$$u \; is \; finite, \qquad T = T_0 \; at \; r = \frac{R(z)}{R_0} \tag{6c}$$

where r and z are the coordinates along radial and axial axes, u and v are the velocity components along the r and z axes respectively, T is the local temperature of the fluid.
Here,

$$\rho_{nf} = (1-\phi)\rho_f + \phi\rho_{f,}\mu_{nf} = \frac{\mu_f}{(1-\phi)^{2.5}}$$
$$(\rho c_p)_{nf} = (1-\phi)(\rho c_p)_f + \phi(\rho c_p)_s, \qquad \alpha_{nf} = \frac{k_{nf}}{(\rho c_p)_{nf}} \tag{7}$$

According to Maxwell [19] and Hamilton and Crosser [20], the values of shape factor of nano particles is as given in the Table1 and also when the nano fluid has 100 times larger thermal conductivity than that of the base fluid, then the thermal conductivity may be expressed as

$$k_{nf} = k_f\left(\frac{k_s + (m+1)k_f - (m+1)(k_f - k_s)\phi}{k_s + (m+1)k_f + \phi(k_f - k_s)}\right) \tag{8}$$

The shape factor m defined in equation (8) is given by $m = \frac{3}{\omega}$ *where* ω is the sphericity of the nanoparticle. The value of sphericity for different shapes of nanoparticles is given in Table 1.
Consider the non-dimensional terms:

$$\bar{r} = \frac{r}{R_0}, \bar{z} = \frac{z}{L_0}, \bar{v} = \frac{L_0}{\delta U}v, \bar{u} = \frac{u}{U}, \bar{d} = \frac{d}{L_0}, \bar{R} = \frac{R}{R_{0,}}$$
$$M^2 = \frac{\sigma B_0^2 R_0^2}{\mu_f}, G_r = \frac{g\alpha R_0^2 T_0 \rho_{nf}}{U\mu_f}, \delta = \frac{\delta}{R_0}, \theta = \frac{T - T_0}{T_0}, \tag{9}$$
$$\bar{p} = \frac{UL_0\mu}{R_0^2}p, \beta = \frac{Q_0 R_0^2}{k_f T_0}, \mu_{nf} = \frac{\mu_f}{(1-\phi)^{2.5}}$$

Here, U is the average velocity along the artery in non-stenotic region. After using (9) in equations (2)-(5) and dropping the dashes,

$$\frac{\partial p}{\partial r} = 0, \tag{10}$$

$$\frac{dp}{dz} = \frac{1}{(1-\phi)^{2.5}}\frac{1}{r}\frac{\partial}{\partial r}\left(r\frac{\partial u}{\partial r}\right) - M^2 u + G_r\theta, \tag{11}$$

$$0 = \frac{1}{r}\frac{\partial}{\partial r}\left(r\frac{\partial \theta}{\partial r}\right) + \beta\left(\frac{k_s + (m+1)k_f + \phi(k_f - k_s)}{k_s + (m+1)k_f - (m+1)(k_f - k_s)\phi}\right) = 0, \tag{12}$$

After non-dimensionalizing the boundary conditions (6a-6c) demonstrating the Figure 1, we get,

$$u = finite \;\; at \; r = 0 \tag{13a}$$

$$\frac{\partial u}{\partial r} = 0, \frac{\partial \theta}{\partial r} = 0 \; at \; r = 0, \tag{13b}$$

$$u = 0 \;\; and \; \theta = 0 \; at \; r = R(z) \tag{13c}$$

and

$$R(z) = \begin{cases} 1 - \dfrac{p^{\frac{p}{(p-1)}}}{p-1}[(z-d) - (z-d)^p]; & d \le z \le d+1 \\ 1, \ otherwise \end{cases} \tag{14}$$

Table1. Shape factor and sphericity for different shaped nanoparticles

S.No.	Name	Shape	Shape factor(m)	Sphericity(ω)
1	Spherical	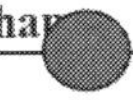	3	1
2.	Blade		8.33	0.36
3	Laminar		16.2	0.185

3. Exact Solutions

Solving equation (12) and using the boundary conditions, exact solution for temperature is obtained as:

$$\theta(r,z) = \left(\frac{R^2 - r^2}{4}\right)\left(\frac{k_f}{k_{nf}}\right)\beta \tag{15}$$

Solving equation (11) and using the boundary conditions, exact solution for velocity is:

$$u = \left(\frac{-G_r\theta + \frac{dp}{dz}}{M^2}\right)\left[1 - \frac{J\left(0, iM\sqrt{(1-\phi)^{2.5}}r\right)}{J\left(0, iM\sqrt{(1-\phi)^{2.5}}R\right)}\right] \tag{16}$$

Flux Q is given by

$$Q = 2\int_0^R ru\, dr \tag{17}$$

$$Q = \left\{\left(\frac{-G_r\theta + \frac{dp}{dz}}{M^2}\right)R^2 - \frac{2\left(\frac{-G_r\theta + \frac{dp}{dz}}{M^2}\right)\left[R^2 HypergeometricPFQ\left[\left\{\frac{1}{2}\right\}, \left\{1, \frac{3}{2}\right\}, \frac{M^2(1-\phi)^{2.5}R^2}{4}\right] - \frac{1}{2}R^2 HypergeometricPFQ\left[\left\{\frac{1}{2}\right\}, \left\{\frac{3}{2}, 2\right\}, \frac{M^2(1-\phi)^{2.5}R^2}{4}\right]\right]}{J\left(0, iM\sqrt{(1-\phi)^{2.5}}R\right)}\right\} \tag{18}$$

Impedance λ is given by

$$\lambda = \frac{\Delta p}{Q}$$

$$\lambda = \frac{\Delta p}{\left\{ \left(\frac{-G_r\theta + \frac{dp}{dz}}{M^2} \right) R^2 - \frac{2\left(\frac{-G_r\theta + \frac{dp}{dz}}{M^2} \right) \left[R^2 HypergeometricPFQ\left[\left\{ \frac{1}{2} \right\}, \left\{ 1, \frac{3}{2} \right\}, \frac{M^2(1-\phi)^{2.5} R^2}{4} \right] - \frac{1}{2} R^2 HypergeometricPFQ\left[\left\{ \frac{1}{2} \right\}, \left\{ \frac{3}{2}, 2 \right\}, \frac{M^2(1-\phi)^{2.5} R^2}{4} \right] \right]}{J\left(0, iM\sqrt{(1-\phi)^{2.5}} R\right)} \right\}} \tag{19}$$

4. Results and Discussion

In this section, we have the results presented through graphs and jotted it down for the crucial flow parameters depicting the nanoparticles' shape effect. Graphs are plotted using MATHEMATICA software. The juxtapositions for these different shapes are placed together and then compared. It is always found that the laminar nano copper particles are better than the spherical nano particles. This study has covered the two extremes of the shape factors range to invoke the shape factor analysis. Through this study we are trying to impart the light on few concepts of structural analysis of nano particles with the hope that this study will contribute to the literature referred by medical practitioners. Flow parameters like velocity, temperature flux and impedance resistance under the influence of Hartmann number M < 1, volume fraction of the nanoparticle (0-0.04) and Grashof number (0.3) have been analyzed.

Figure 2 is the variation of flux along radial axis r for different values of Hartmann number M and for different nanoparticles. The graph indicates that the flux is minimum at r = 0 and it increases gradually as r increases for all values of Hartmann number. It is also seen that for a given Hartmann number, Flux is more in the case of laminar nanoparticles other than two different nanoparticles.
Figure 3 shows the change in impedance resistance for different values of z and Hartmann number M. It decreases as the value of z increases. It can also be noticed that the impedance resistance is more for spherical nanoparticles and the laminar has least for a given value of Hartmann number M.

Velocity profile u(r, z) against the radial axis r for different values of the nanoparticle volume fraction ϕ is plotted in Figure 4(a). This graph shows that the velocity profile is symmetric about $r = 0$. Velocity increases as the value of r approaches to zero and it decreases as r moves away from zero as shown in Figure 5(a). With the increase of the value of nanoparticle volume fraction, the velocity profile decreases. Velocity is high for the lower values of nanoparticle volume fraction. It is also found that the velocity is high for laminar nanoparticles than the other two shapes.

In Figure 4(b), velocity profile is obtained for three different types of nanoparticles- laminar nanoparticles, blade nanoparticles and spherical nanoparticles. It is found from the graph that the velocity is high for laminar nanoparticles compared to blade and spherical nanoparticles for a given nanoparticle volume fraction.

The effect of spherical, blade and laminar nanoparticles on the temperature profile against the radial axis r is shown in Figure 5(a). From this figure it is noticed that the temperature is maximum at$r = 0$and it decreases as the radius increases. Figure 5(a) shows that the temperature increases rapidly when there is a rise in absorption parameter β and it is also found that the temperature is high in case of spherical nanoparticles compared to blade and laminar nanoparticles for a fixed value of β.

Figure 5(b) depicts the temperature profile for different values of stenosis height δ. It can be seen that the temperature profile is symmetric about the axis $r = 0$. The value of temperature increases as the height of the stenosis increases for all the types of nanoparticles considered. It is also found that the temperature is high for spherical nanoparticles than that of blade and laminar nanoparticles.

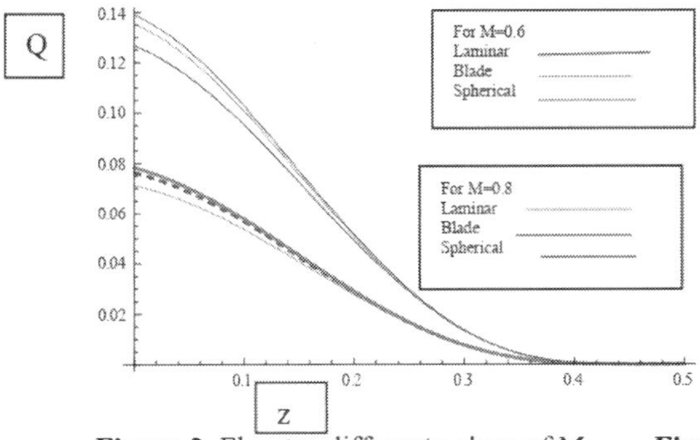

Figure 2. Flux for different values of M.

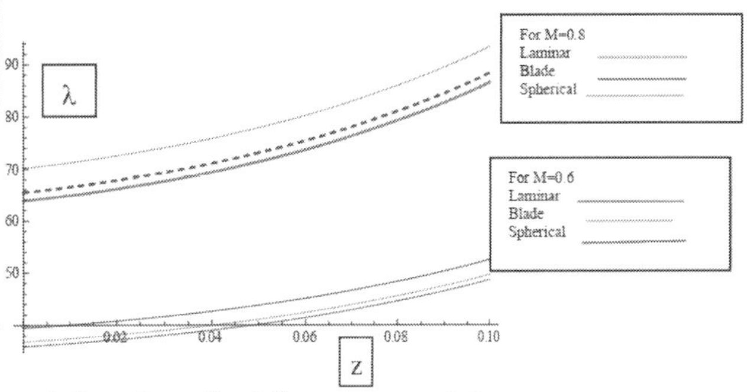

Figure 3. Impedance for different nano particles

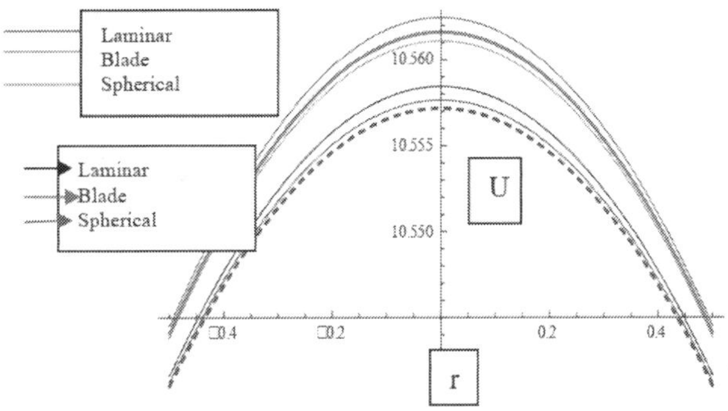

Figure 4(a)

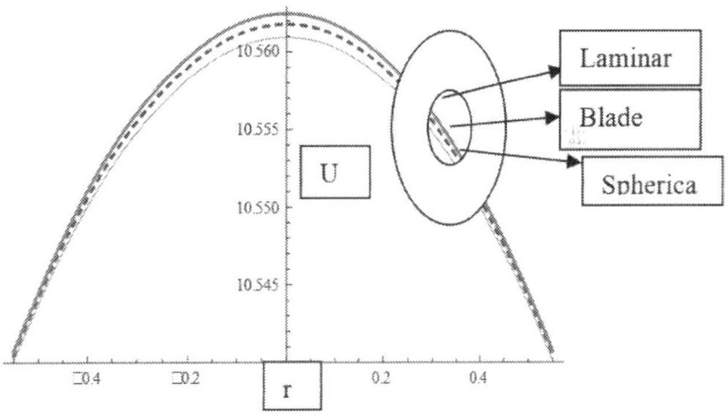

Figure 4(b)

Figure 4. Velocity profile against r for (a) different nanoparticle volume fraction (b) same nanoparticle volume fraction for spherical, blade and laminar nanoparticles.

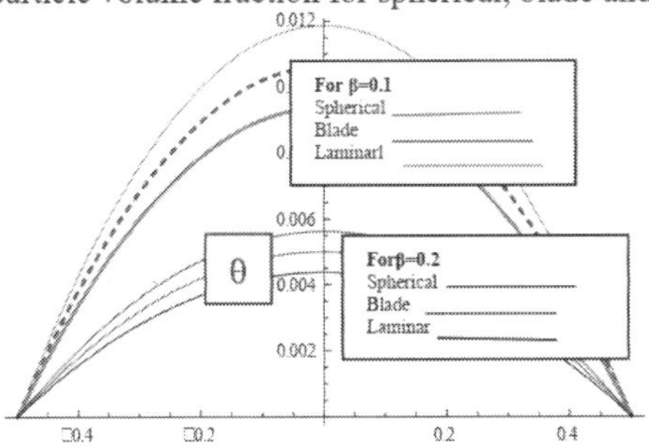

Figure 5(a)

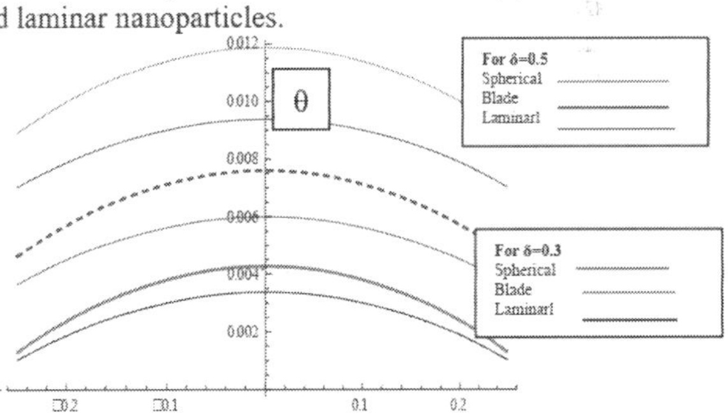

Figure 5(b)

Figure 5. Temperature Profile for (a) laminar, blade and spherical nanoparticles and (b) different values of stenosis height δ.

5. Conclusion

In this manuscript, the focused nanoparticle is copper with water as base fluid and their properties are defined using Hamilton and Crosser model. This study is mainly focused on the two extremes of nano shape factor and tried imparting the knowledge related to some significant parameters which are directly related to medical interest. Among the different shaped nanoparticles considered in Table1, spherical nanoparticles have the least shape factor (3.0) and laminar nanoparticles have the highest shape factor (16.2). Hence, study of these two kinds of nanoparticles is important to analyze the characteristics which vary depending on the shape factor m. The flow governing equations have been solved analytically and using Mathematica. The important observations of the current work are:

- Velocity of the nano particles in the stenosed artery derived mathematically and plotted in the graphs against center of line r. Laminar nanoparticles may be used for treating the stenosed artery as it has velocity profile better than blade and spherical shaped copper nanoparticle.
- Temperature is high for Spherical nanoparticles compared to Blade and Laminar nanoparticles for a fixed value of β. The temperature is inversely proportional to the thermal conductivity. Spherical particles are more conductive as compared to both of the other shape for the same value of volume fraction.
- Temperature decreases rapidly when there is a rise in the heat absorption parameter β for the volume fraction φ=0 to 0.5.
- Flux decreases as the Hartmann number M increases. It can be concluded from the Fig4 that laminar nano particles in presence of magnetic field gives a better flux as compared to blade and spherical nano particles. Improved flux with the laminar shape confirms that the stenosed area can be treated faster with laminar shaped nanoparticles.
- Impedance resistance is high at the center line and it is high in the case of spherical as compared to laminar nanoparticles.
- Velocity profile is symmetric about center line $r = 0$ and it attains its maximum at center $r = 0$. Velocity is less in the case of spherical nanoparticles than that in laminar nanoparticles.

The study may be extended considering different nanoparticles and their effects on various flow parameters.

Nomenclature

Symbol	Meaning
ρ_{nf}	Effective density
μ_{nf}	Effective dynamic viscosity
$(\rho_{cp})_{nf}$	Heat capacitance
α_{nf}	Effective thermal diffusivity
k_{nf}	Effective thermal conductivity
ϕ	Solid nanoparticle volume fraction
M	Hartmann number
β	Heat absorption parameter
G_r	Grashof number

References

[1] E. E. Tzirtzilakisa, A mathematical model for blood flow in magnetic field, PHYSICS OF FLUIDS. 2005;17, 077103.

[2] R Ponalagusamy, Blood Flow Through an Artery with Mild Stenosis: A Two-Layered Model, Different Shapes of Stenoses and Slip Velocity at the Wall, Journal of Applied Sciences. 2007; 7(7); 1071-1077.

[3] S.U. Siddiqui, N.K. Verma, Shailesh Mishra, R.S. Gupta, Mathematical modelling of pulsatile flow of Casson's fluid in arterial stenosis, Applied Mathematics and Computation. 2009;

210.
[4] Gaurav Varshney, V.K. Katiyar, Sushil Kumar, Effect of magnetic field on the blood flow in artery having multiple stenosis: a numerical study, International Journal of Engineering, Science and Technology. 2010; Vol. 2, No. 2.
[5] J. Venkatesan, D. S. Sankar, K. Hemalatha, and Yazariah Yatim, Mathematical Analysis of Casson Fluid Model for Blood Rheology in Stenosed Narrow Arteries, Journal of Applied Mathematics. 2013;
[6] Srivastava, Neetu, Analysis of Flow Characteristics of the Blood Flowing through an Inclined Tapered Porous Artery with Mild Stenosis under the Influence of an Inclined Magnetic Field, Journal of Biophysics, Volume 2014, Article ID 797142.
[7] Srivastava,NeetuThe Casson fluid model for blood flow through an inclined tapered artery of an accelerated body in the presence of magnetic field,Int. J. Biomedical Engineering and Technology. 2014; Vol. 15, No. 3.
[8] Noreen Sher Akbar, S.U. Rahman, R. Ellahi, S. Nadeem, Nano fluid flow in tapering stenosed arteries with permeable walls", International Journal of Thermal Sciences. 2015; 85.
[9] Ondrej Vesely, Ludmila Novakova, Josef Adamec, Experimental Study of effect of Stenosis Geometry on flow parameters", EPJ Web of Conferences. 2015; 0210.
[10] Noreen Sher Akbar, and M. T. Mustafa, Ferromagnetic effects for nanofluid venture through composite permeable stenosed arteries with different nanosize particles, AIP Advances. 2015; 5, 077102.
[11] Noreen Sher Akbar, Adil Wahid Butt, Magnetic field effects for copper suspended nanofluid venture through composite stenosed arteries with permeable wall, Journal of Magnetism and Magnetic Materials. 2015.
[12] Ashfaq Ahmed ,Sohail Nadeem, Shape effect of Cu-nanoparticles in unsteady flow through curved artery with catheterized stenosis, Results in Physics 7. 2017; 677-689.
[13] Asha K N and Neetu Srivastava, Magnetic repercussion on the different shaped copper nanoparticles to improve vasodilatation effect in stenosed artery, Materials Today, Volume 44, Part 1, 2021, Pages 2142-2146
[14] A. Nematollahzadeh, A. Dabaleh, N. Ahadi-Jomairan, S. Torabi, Iron-oxide nano-particles effect on the blood hemodynamic in atherosclerotic coronary arteries", Chemical Engineering Science. 2017.
[15] DzulianaFatin Jamil, RozainiRoslana, Mohammed Abdulhameedb,Norziha Che-Hima, SuliadiSufahania, Mahathir Mohamada, Muhamad Ghazali Kamardana, Unsteady Blood Flow with Nanoparticles Through Stenosed Arteries in the Presence of Periodic Body Acceleration, IOP Conf. Series: Journal of Physics. 2018; Conf. Series 995.
[16] Asha K N and Neetu Srivastava, Geometry of stenosis and its effects on theblood flow through an artery - A theoretical study, AIP Conference Proceedings 2375, 030006 (2021);
[17] K Gayathri, K Shailendhra, MRI and Blood Flow in Human Arteries: Are There Any Adverse Effects? Cardiovasc Eng Technol. 2019 Jun;10(2):242-256. doi: 10.1007/s13239-019-00400-x. Epub 2019 Jan 28.
[18] Arjun Suresh and Vikas Rajan, Study of non-Newtonian blood flow through arteries using Open FOAM, AIP Conference Proceedings 2134, 040003 (2019); https://doi.org/10.1063/1.5120211
[19] J. C. Maxwell, A Treatise on Electricity and Magnetism, Clarendon Press. 1873; Oxford, Vol. 1.
[20] R.L. Hamilton and O.K.Crosser, Thermal Conductivity of heterogeneous two-component system", Industrial and Engineering Chemistry Fundamentals 1. 1962

Design and Analysis of a Compact Hydraulic Bolt Tensioner

Chethan Kumar G[1], Srinivas Reddy Mungara[2] and Joseph Sanjan[3]

[1] Assistant Professor, Department of Mechanical Engineering, CMR Institute of Technology, Bengaluru 560037.

[2] Assistant Professor, Department of Mechanical Engineering, School of Engineering and Technology, CMR University, Bengaluru 562149.

[3] Assistant Professor, Department of Mechanical Engineering, CMR Institute of Technology, Bengaluru 560037.

Chethankumarg01@gmail.com

Abstract. In equipment like high pressure vessels, there will caps on which loads of the order of several thousands of tons acts. Such caps require a huge number of big-diameter bolts, often M70 to M150mm in diameter. Each bolt will be subjected to a tremendous amount of force, on the order of 300 tons. The use of hand-operated wrenches to apply such a load is deemed out. When a high number of bolts are tightened on the flange, the applied load is lost owing to friction at the clamping surface and the applied load is not uniform. It is advised that a hydraulic bolt tensioning system be used to address these drawbacks. The device operates at a pressure of 1500 bar, which is extremely high. Therefore, the equipment must be designed to ensure the device's compactness and safety. Bolt tensioners are used to tighten bolts and studs in a variety of important applications. These devices would be imported. It was proposed in this work that a small hydraulic bolt tensioner be designed and analysed using both theoretical and finite element analysis (Non-liner static analysis). The required boundary conditions and applied pressure were computed. The tensioner's design was optimized to assure its safety and compactness.

1. Introduction

Hydraulic Bolt Tensioner (HBT) is an equipment for tightening bolts and studs. Because of several advantages such as ease of use, great accuracy, repeatability, light weight, portability and the replacement of huge mechanical effort to create torque with hydraulic pressure, HBT are employed in a wide range of applications [1]. Pipeline flanges, heat exchangers, pressure vessels, compressor covers, boiler feed pumps, and windmills are indeed examples of areas.

A bolt tensioner is necessary to tighten the bolt and nut; hence, considerable mechanical effort is required to create torque. The direct axial stretching of the bolt to create preload is known as bolt tensioning. Also, while tightening a large number of bolts on the flange, it is difficult to maintain consistent torque owing to friction load losses at the clamping surface. To address the aforementioned limitation, HBT are employed to tighten the bolts and studs. HBT is comprised of a hydraulic load cell and a threaded puller as shown in figure 1. Hydraulic load cell uses piston and cylinder arrangement. The piston is placed in an elastic diaphragm. The piston doesn't come in contact with the load cell. When a weight is applied, the piston moves, and the threaded puller moves with it.

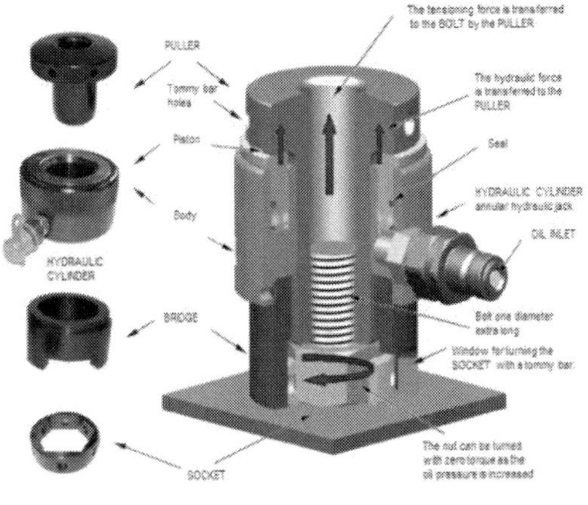

Figure 1. HBT

The most significant aspect in the operation of HBT is the effective tensile coefficient, which is the ratio of required clamping force to initial tension applied by a hydraulic tensioner. In one particular research, the tightening process was investigated using the finite element approach for axisymmetric elastic contact. The coefficient values were evaluated for several design cues on surface contact [2]. The effective tensile coefficient was computed using a simple and comprehensive technique, and the magnitude of the coefficient was evaluated using a simple and practical equation. In addition, with the influence of the latched plate's young's modulus included, the correlation between effective tensile coefficient and grasp length was explored, and the obtained outcomes were compared using experiments and the FEM technique [3].An investigation was conducted on the technique for controlling and evaluating the joint distortion and strain fields of the tightened fastener, and to develop a reliable interdependency with the initial clamp load, where initial clamp load is vital in the process of bolt tightening [4].

A set of principles for the joint creator, specifier, or apparatus selector to ensure that the best and most reasonable system is chosen for each joint has been detailed in one research [5], however compact tensioner and frictional losses have not been explored. Another study [6] looked at the total amount of overload required to minimize load loss in a tensioner, but the compact design of tensioners and frictional losses were not discussed. Though load loss due to nut to tensioner interaction and bolt to bolt interaction has been discussed in detail. Large diameter bolts, particularly for the assembly of low strength bolts, have a practical application and hence, has been taken into consideration in this work. Though a hydraulic bolt tensioner is already in use, issues such as friction loss at the clamping surface, non-uniformity of the applied load when a large number of bolts are tightened on the flange, and the massive mechanical effort required to tighten such large bolts and studs, which is difficult, are yet to be researched. As the equipment operates at a very high pressure, which is nearly 1500bar; hence, the device was designed to ensure its compactness and safety in this work.

2. Methodology

The hydraulic bolt tensioner must first be designed [7]. A number of iterations were performed to develop the bolt tensioner, as well as design optimization to achieve optimum stress levels in the bolt tensioner by varying the design characteristics for theoretical examination as well as finite elemental analysis. Followed by estimation of circumferential and radial stress level in hydraulic cylinder to depict the safety through number of iteration and also failure of bolt in different trails. Table 1 shows the characteristics of bolt tensioner selected for the design. Force the selected pressure and force, hydraulic pressure is estimated to be 20000mm^2.

Table 1. Characteristics of Bolt tensioner

Characteristics	Values
Maximum bolt load	30000N
Bolt size	M100
Maximum hydraulic pressure applied	1500MPa
Weight of the device	30 Kg
Material	High strength stainless steel
Young's modulus, E	210000MPa
Density ρ	7800kg/mm3
Force	30000 N
Pressure Applied	150 MPa

3. Design and Finite Elemental Analysis of bolt Tensioner

Multiple iterations were carried out to reach the optimal stress level in the bolt tensioner, including altering the cylinder diameter both internally and externally, inserting fillet, and so on.

3.1 Iteration 1

The outer radius of the hydraulic cylinder {R} is adjusted to 102 mm^2 and the inner radius {r} to 63 mm^2 during first iteration. The hydraulic cylinder's total area is estimated to be 20216.1mm^2. As a result, the estimated weight of the cylinder has been revised to 30.3 kg. To design of bolt tensioner theoretical calculations has been carried out by considering the characteristics selected for the analysis. In addition to this bolt dimensions, flange area has been considered as shown in figure 2. The results revealed that the imposed pressure causes extremely high stress levels. The equipment cannot be used at these levels of stress. As a result, stress levels must be reduced to a minimum.

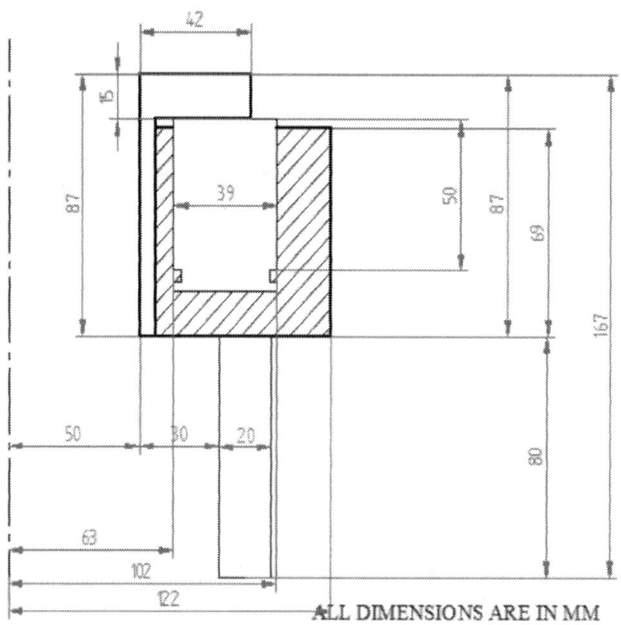

Figure 2. Design dimension of Bolt Tensioner

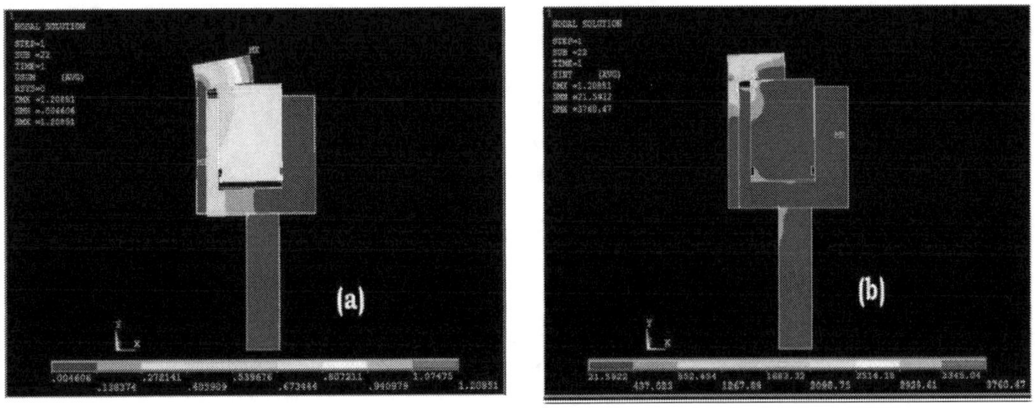

Figure 3. (a) Deformation and (b) Stress of bolt tensioner.

Ansys software creates an axi-symmetric tensioner model for the aforementioned dimensions using the solidquad44 element. The model was meshed, and boundary conditions and a pressure of 150MPa were imposed. A nonlinear contact analysis was performed on a bolt tensioner finite element model. The resulting deformation and tension were depicted in figure 3 (a) and (b).

3.2 Iteration 2

For the second iteration, a fillet with a radius of 3mm was introduced to the puller where the stress level was high, and a fillet with a radius of 2mm was inserted at the upper corners of the piston, as illustrated in figure 4. The findings of finite element analysis produced in the second iteration were lower than those obtained in the first iteration, as illustrated in figures 5 (a) and (b). The obtained stress levels remained high at the end of the second repetition. Though the achieved stress levels were not safe for device operation, design improvements were made in order to attain optimal stress levels.

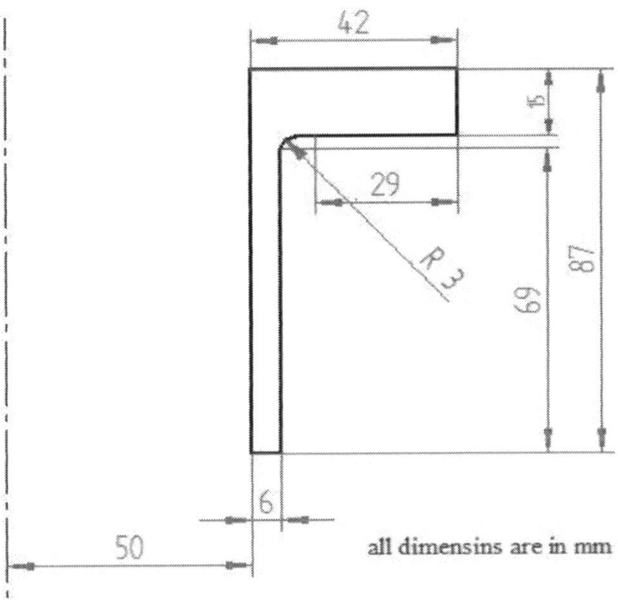

Figure 4. Dimension of puller in Bolt Tensioner

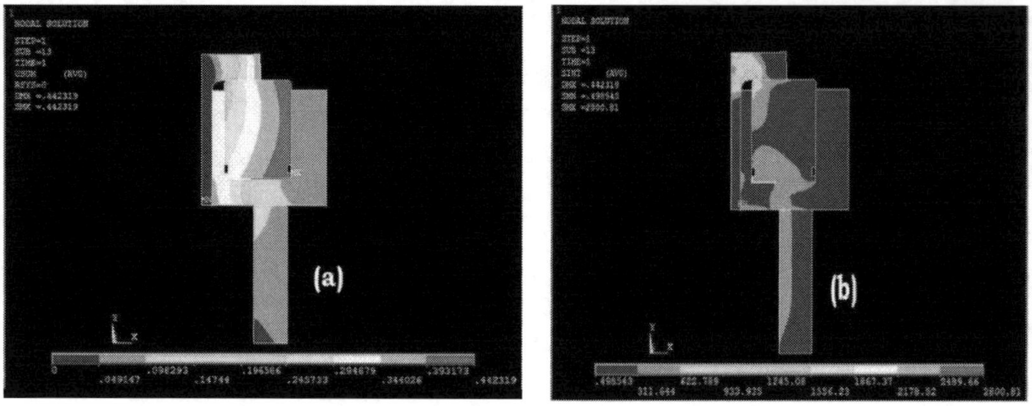

Figure 5. (a) Deformation and (b) Stress of bolt tensioner.

3.3 Iteration 3

Puller dimensions has been changed and also new dimension were introduced to bolt tensioner for the third iteration as shown in figures 6 (a) and (b). For the third iteration, outer radius of hydraulic cylinder has been changed to 106 mm considering the area estimated in 1st iteration. Hence, inner diameter estimated to be 69.28 mm. The findings of finite element analysis produced in the third iteration were lower than those obtained in the first iteration and second, as illustrated in figures 7 (a) and (b). Stress levels were lower compared to the preceding two iterations, and deformation was relatively minimal. However, because the resulting stress levels were too high to safely run the device, iteration was performed to lessen the stress in the bolt tensioner.

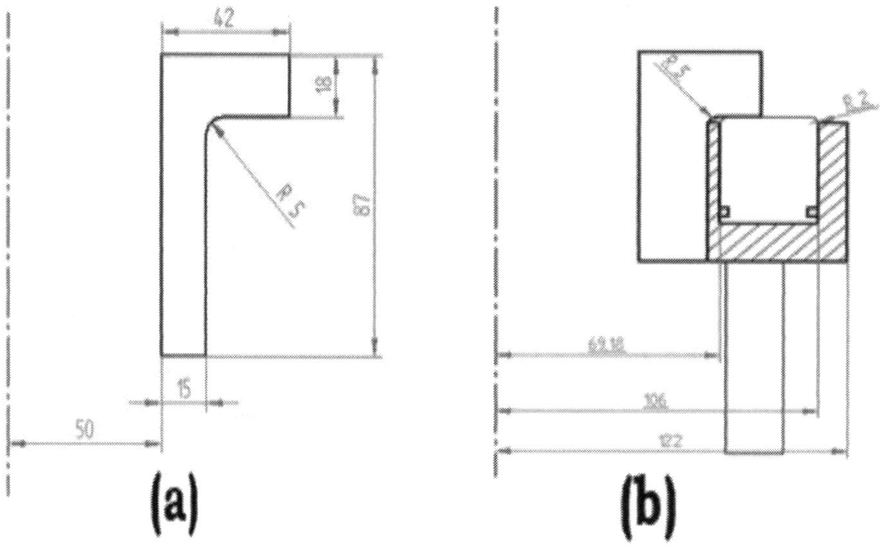

Figure 6. (a) Threaded puller and (b) new dimension of bolt tensioner

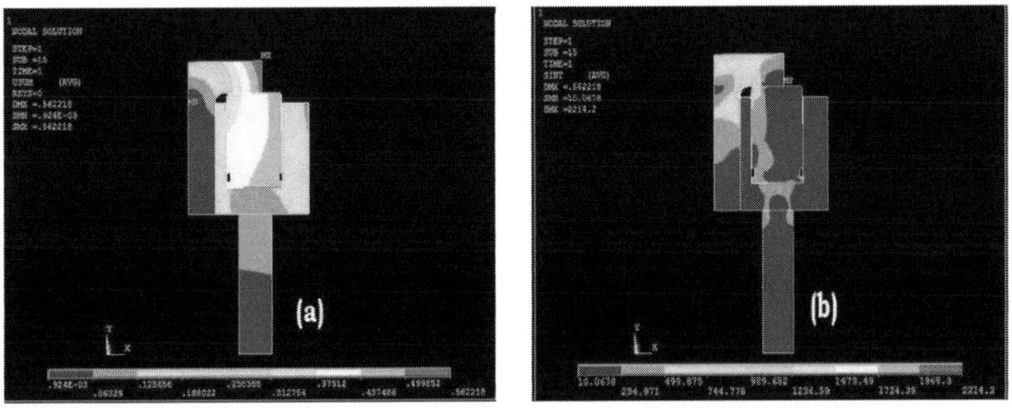

Figure 7. (a) Deformation and (b) Stress of bolt tensioner.

3.4 Iteration 4

Puller dimensions has been changed again, also new dimension was introduced to bolt tensioner and piston as shown in figures 8 (a) (b) and (c). The findings of finite element analysis produced in the fourth iteration were lower than those obtained in the first iteration and second, as illustrated in figures 9 (a) and (b).

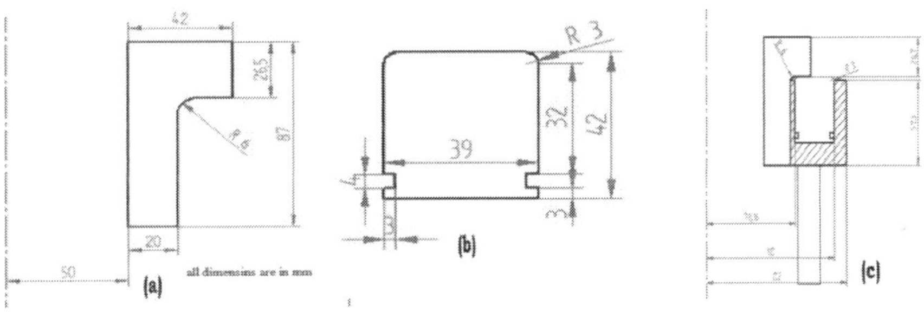

Figure 8. (a) Diagram of threaded puller, (b) Diagram of piston and (c) Dimension of bolt tensioner

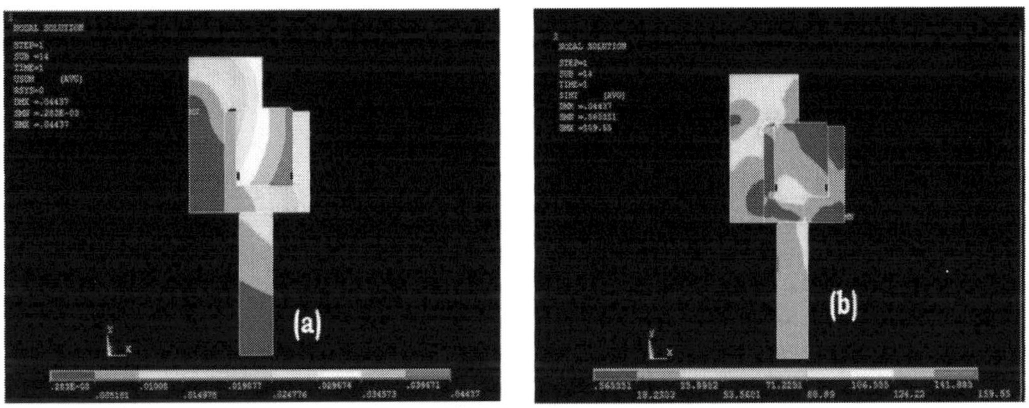

Figure 9. (a) Deformation and (b) Stress of bolt tensioner.

To ensure the compactness and safety of a hydraulic bolt tensioner, four iterations were carried out by changing dimensions of a puller, piston and hydraulic cylinder to minimize the stress levels. In first iteration the stress obtained was very high to 3760MPa. To reduce the stress level, the design optimization was carried out and at the end of the forth iteration the stress levels was found to be minimum which was 159MPa.Accordingly the deformation was also minimum as shown in the table 2. For the applied pressure of 150MPa the obtained stress is minimum and safe to operate.

Table 2. Stress and Deformation of Bolt tensioner

Number of iterations	Stress (MPa)	Deformation (mm)
1	3760	1.2
2	2800	0.44
3	2214	0.56
4	159	0.44

4. Finite Elemental and Theoretical analysis of Hydraulic Cylinder

The primary goal of this study is to assure the safety and compactness of a hydraulic bolt tensioner that can function at high pressures of 150MPa. The hydraulic cylinder was subjected to pressure; the stresses formed in the cylinder must be evaluated. To compare the stresses created in the cylinder, a finite element analysis was performed. The inner and outer radius of the cylinder was 102mm and 122mm, respectively, in the first iteration. Circumferential and radial stresses induced in cylinder are shown in figure 10 (a) and (b) respectively.

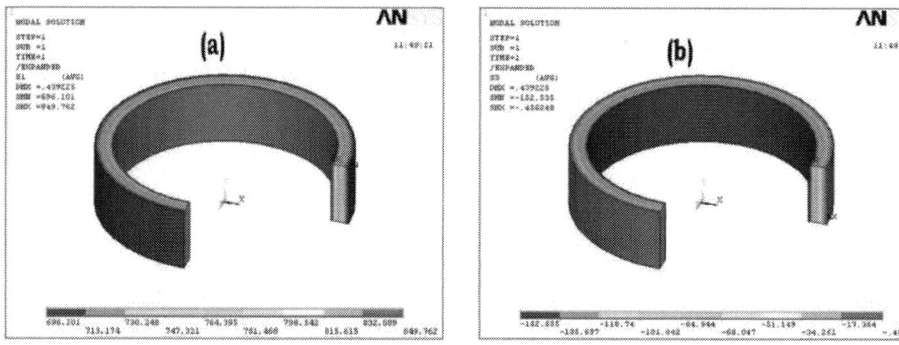

Figure 10. (a) Circumferential stresses and (b) Radial stresses induced in the cylinder in first and second iteration.

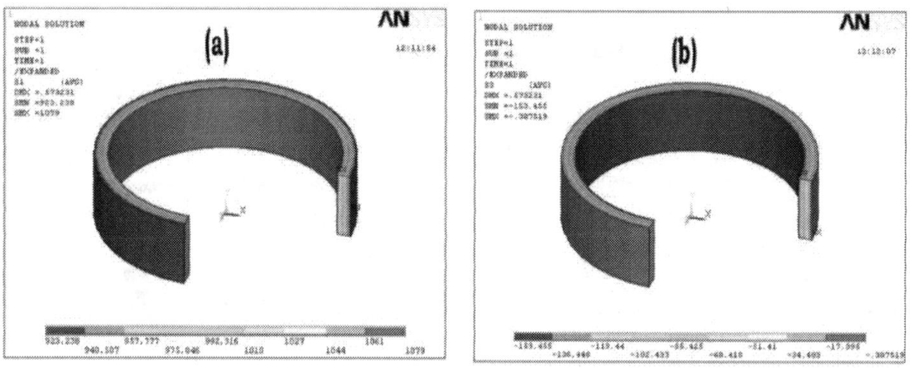

Figure 11. (a) Circumferential stresses and (b) Radial stresses induced in the cylinder in third iteration.

The cylinder dimensions were n ot changed in the second iteration. Similar to the first iteration, the stresses created in the cylinder were consistent. The radial stress was -152 MPa and the circumferential stress was 849 MPa as shown in figure 10 (a) and (b).

The inner and outer radius of the cylinder was altered to 106 mm and 122 mm, respectively, in the third iteration. The results obtained are shown in figure 11 (a) and (b). The stresses induced in the cylinder are high so iteration is carried out get optimum stress levels. In forth iteration the inner and outer radius of the cylinder are changed to 112 mm to 122 mm. Finite element model of cylinder is modelled with the same material properties. The results obtained as shown in figure 12 (a) and (b)

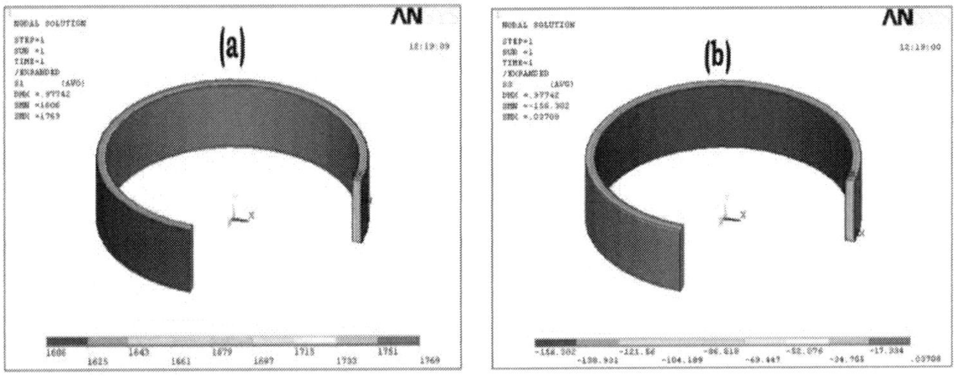

Figure 12. (a) Circumferential stresses and (b) Radial stresses induced in the cylinder in fourth iteration.

The hydraulic cylinders thickness is 1/20th of its diameter. As a result, the hydraulic cylinder is referred to as a thick cylinder. The stresses were calculated using the following equation [8] for a thick cylinder.

$$\text{Radial Stress, } \sigma_r = \frac{p_a\, a^2 - p_b\, b^2}{b^2 - a^2} - \frac{a^2 b^2}{r^2}\frac{p_a - p_b}{b^2 - a^2} \quad (1)$$

$$\text{Circumferential Stress, } \sigma_\theta = \frac{p_a\, a^2 - p_b\, b^2}{b^2 - a^2} + \frac{a^2 b^2}{r^2}\frac{p_a - p_b}{b^2 - a^2} \quad (2)$$

When thick cylinder which is subjected to internal pressure, p_b=0 and p_a = r. The dimensions of the hydraulic cylinder in the second iteration were equivalent to those in the first iteration, and so the stresses induced in the cylinder were equivalent to those in the first iteration. Table 3 illustrates the theoretically calculated radial and circumferential stress using equation 1 and 2 induced by changing the cylinder's dimension in various iterations.

Table 3. Radial and Circumferential stresses of Bolt tensioner

Iterations	Radial Stress (MPa)	Circumferential Stress (MPa)
1	-150	846.89
2	-150	846.89
3	-150	1072
4	-150	1753

Table 4 illustrates the circumferential and radial stress induced in various iterations as estimated using finite element analysis and theoretical analysis. Stresses predicted by theoretical and finite elemental analysis were observed to be in good accordance.

Table 4. Results of Theoretical and finite elemental analysis.

Iterations	Circumferential stress (σ_θ) (MPa)		Radial stress (σ_r) (MPa)	
	Theoretical Analysis	Finite Elemental Analysis	Theoretical Analysis	Finite Elemental Analysis
1	847	849	-150	-152
2	847	849	-150	-152
3	1072	1079	-150	-153
4	1753	1769	-150	-159

5. Finite Element and Theoretical Analysis of Bolt Failure

There is a risk of thread wear and bolt failure if the puller lifts the bolt during the tightening operation. To avoid this, a study of bolt failure was performed for M100 diameter bolts with the dimensions presented in figure 13. For the safety of bolt, 2 different trails were carried out considering the length of 167mm and 72mm respectively. In trail 1, M100 bolt of length 167mm was modelled in Ansys and estimated the deformation. Similarly, in trail 2, bolt length was reduced to 72mm to reduce the deformation induced in trail 1. Figure 14 (a) and (b) shows the deformation of bolt length 167mm and 72mm respectively.

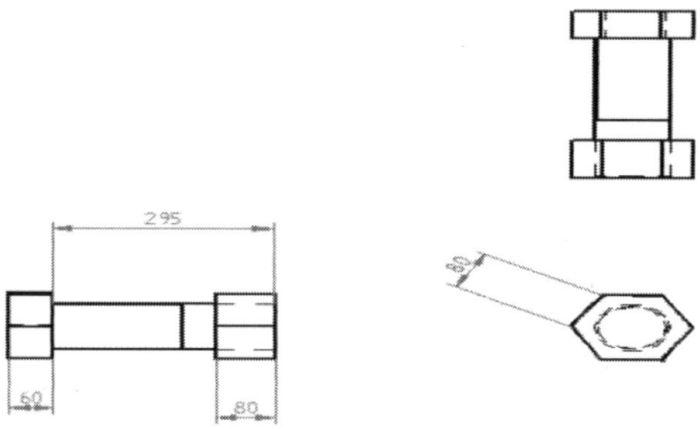

Figure 13. Diagram of M100 bolt and nut

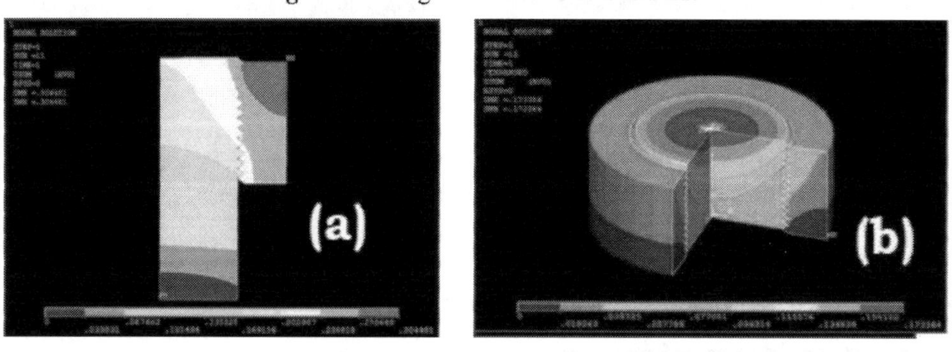

Figure 14. Deformation of bolt length (a) 167mm and (b) 72mm respectively.

Table 5. Features of Bolt Employed for the Theoretical Estimation

Sl No.	Characteristics	Values
1	Diameter of bolt	100 mm
2	Force	30000 N
3	Area of bolt	7854 mm^2
4	Pressure	38.19 MPa

The features of the bolt employed for the theoretical estimation are shown in Table 5. For the analysis, the same features were used as in the finite elemental investigation. For the diameter of the bolt 100mm and the applied force of 30kN, area and pressure has been calculated using equation 3. Deformation has been estimated using equation 4 in 2 trails considering length of the bolt as 167mm and 72mm respectively. Table 6 illustrates the deformation estimated in 2 trails compared with Finite Elemental Analysis.

$$\text{Pressure}=\frac{\text{Force}}{\text{Area}} \quad (3)$$

$$\text{Deformation, } \delta=\frac{\text{pl}}{\text{AE}} \quad (4)$$

Table 6. Comparison of Deformation Estimated Using Finite Elemental Method

Length of Bolt (mm)	Theoretical Analysis (mm)	Finite Element Analysis (mm)
167	0.3	0.3
72	0.17	0.17

From bolt failure results, it was observed that the deformations are very negligible so that the chances of failure and thread wear are minimum when the puller stretches the bolt. The design optimization the tensioner carried to optimize the stress levels. In fourth iteration the stress levels are found to be safer to operate the tensioner and the fourth iteration dimensions may be proposed to design the bolt tensioner. Also, 72mm length of the bolt could be selected since deformation is very small. Hence, chances of bolt failure are minimum.

6. Conclusion

A tensioner's design has been optimised to reduce stress levels and assure its safety and compactness. The stress levels are 3760MPa in the first iteration, which is very high, and 159MPa at the conclusion of the fourth iteration, which is quite low. The fourth iteration's stress levels are confirmed to be safe. Both the Finite Elemental technique and analytical methods produce comparable stresses in the hydraulic cylinder. The fourth iteration dimensions are provided for hydraulic bolt tensioner design based on analysis and verification. To assure the safety of the bolt, the Finite Elemental technique and analytical method are used to verify bolt failure. Thread wear and bolt failure are determined to be highly improbable. Direct loading also results in less load losses.

References

[1] Monville, J.-M. (2016). Optimal tightening process of bolted joints. International Journal for Simulation and Multidisciplinary Design Optimization, 7, A4.

[2] Fukuoka, T. (November 1, 1992). "Finite Element Simulation of Tightening Process of Bolted Joint with a Tensioner." ASME. J. Pressure Vessel Technol. November 1992; 114(4): 433–438.

[3] Fukuoka, T. (November 1, 1994). "Analysis of the Tightening Process of Bolted Joint With a Tensioner Using Spring Elements." ASME. J. Pressure Vessel Technol. November 1994; 116(4): 443–448

[4] Nassar, S. A., &Meng, A. (2007). Optical Monitoring of Bolt Tightening Using 3D Electronic Speckle Pattern Interferometry. Journal of Pressure Vessel Technology, 129(1), 89. doi:10.1115/1.2389024

[5] Nassar, S. A., and Meng, A. (January 4, 2006). "Optical Monitoring of Bolt Tightening Using 3D Electronic Speckle Pattern Interferometry." ASME. J. Pressure Vessel Technol. February 2007; 129(1): 89–95.

[6] Brown, W. Hydraulic Tensioner Assembly: Load Loss Factors and Target Stress Limits. Proceedings of the ASME 2014 Pressure Vessels and Piping Conference. Volume 3: Design and Analysis. Anaheim, California, USA. July 20–24, 2014. V003T03A018. ASME.

[7] Yildirim, Nihat. (2014). A Study on Comparison of Bolt Tensioning Applied by Conventional Torque Method and Hydraulic Tension Method.

[8] Hassan, Sachidananda. (2019). Design and Analysis of Pressure Vessel. International Journal of Mechanical and Production Engineering Research and Development, Vol. 9, Issue 5, Oct 2019, 125–

Temperature Effect On Reactive Powder Concrete Using Sillimanite As Fine Aggregate

Sophia. M[1], Soundarya N[2]
[1]Assistant Professor, Department of Civil Engineering, CMR Institute Technology, Bengaluru, India
[2]Associate Professor, Department of Civil Engineering, CMR Institute of Technology, Bengaluru, India

soundarya.n@cmrit.ac.in

Abstract. This study examines the use of sillimanite as a fine aggregate in the production of temperature-resistant reactive powder concrete. The impact of high temperatures on the mechanical strength of reactive powder concrete with High Alumina cement is investigated. The effectiveness of utilizing glass powder and sillimanite on the mechanical properties of RPC at high temperatures is investigated in this research. The samples were heated in the muffle furnace to the desired temperatures and then tested for their residual compressive strength, flexural strength and split tensile strength. The residual values of compressive strength, flexural strength and split tensile strength were measured at the temperature of 270C to 800°C. The weight loss of the specimens after exposure to the elevated temperatures was measured and the values showed enhanced resistance to the high temperature effects. The results demonstrate the greater contribution of glass powder and sillimanite towards the significant improvement of high temperature strength of reactive powder concrete than those made with normal quartz sand as fine aggregate.

1. Introduction

A Fire is regarded as one of the serious threat for concrete construction. Recent advances in the concrete technology have yielded high strength concrete with strength values reaching multi-fold times of normal concrete [1]. This ultra-high strength is usually attained by employing various techniques such as exclusion of coarse aggregates, inclusion of high dosage of cement, super plasticizer, silica fume and steel fibres [2, 3]. More recent developments in concrete technology mainly focussed on improving the strength performance and durability of concrete by replacing the conventional concrete materials to yield high strength [4]. Industrial by-products and extra cementitious materials are increasingly being used in concrete for this purpose [5, 6]. RPC is a very promising building material that finds applications in all fields of civil engineering due to their outstanding mechanical properties and extreme toughness [7, 8]. Several studies have already been done to further improve the super-high performance of RPC to arrive with the new generation of concrete with ultrahigh strength, remarkable durability and extreme toughness [9, 10]. This present study attempts to utilize sillimanite as a partial substitute of fine aggregate. The quartz powder which is the commonly used fine aggregate for stabilising the RPC is replaced by about 50% using sillimanite in this study. Several attempts to use waste materials and environmentally friendly products in the building sector are emerging in recent years, with the goal of establishing eco-technology in the construction area [11, 12]. Reactive powder concrete due to its ultra high strength and extreme packing density undergoes severe spalling due to fire [13, 14]. Hence the study of fire properties of Reactive powder concrete makes it mandatory before it can be employed for practical civil engineering applications.

With detailed survey of recent research works very limited works have been done on the mechanical properties of Reactive powder concrete after exposure to fire. The aim of this present work is to study the effect of using glass powder and sillimanite on the performance of Reactive powder concrete after exposure to fire. The mechanical properties such as compressive strength, flexural strength and split tensile strength have been experimentally investigated after exposure to high temperature range (20°C to 900°C). Furthermore, the pore sizes of the RPC mixes were studied with the help of mercury intrusion Porosimetry.

2. Materials and Methodology

The commercially available high alumina cement and micro silica fume were used as main binder for the production of Reactive powder concrete. The quartz sand, sillimanite and waste glass powders were used as fine aggregate. The brass coated steel fibres and liquid polycarboxylate based super plasticizer were used. The detailed mix design of Reactive powder concrete is given in table 1.

Table 1 Mix proportion of RPC mix

Mix	Cement (%)	Silica fume (%)	Fine aggregate (%)			Steel fibre (%)	SP
			Quartz sand	Sillimanite	Glass powder		
Control	1	-	1	-	-	2%	2%
Mix 1	1	0.3	0.4	0.3	0.3	2%	2%

The reactive powder concrete specimens were prepared by initially dry mixing the materials for about 5 minutes after which the required amount of water and super plasticizer were added to the premixed component. The mix was continuously mixed to achieve better consistency and was cast in 50mm cubic specimens for compressive strength test and 40x40x160 mm prismatic specimens for flexural strength. For split tensile test cylindrical specimen of diameter 50mm and height 100mm were cast. Both Compressive strength specimen and split tensile test specimen is shown in the figure 1. The specimens were de-moulded after water for curing up to 28 days. The samples were heated at 100°C, 300°C, 400°C, 500°C, 700°C and 900°C as target temperatures using muffle furnace and were maintained for about 1 hour to ensure a steady state temperature within the concrete at the target temperature. The specimens were cooled naturally to attain the ambient temperature before testing compressive strength, flexural strength and split tensile strength. The SEM pictures were taken using a ZEISS EVO SEM in secondary electron mode at a 40 kV accelerating voltage. Through line analysis methods, the EDAX analysis was also done to examine the active ingredients contained in the concrete mixes.

Figure 1 Compressive and Split Tensile Strength test specimen

3. Results and Discussions

3.1 Residual Compressive Strength

The residual compressive strength of Reactive powder concrete is shown in figure 2. The strength of the concrete mixes showed increased strength after exposure to 300°C and attained compressive strength was 28% more than the strength of the normal temperature. This improvement in the compressive strength may be due to the hardening of Reactive powder concrete specimens due to the progress of internal hydration reaction caused by the heated vapours inside the concrete [15]. However above 500°C the residual compressive strength of Reactive powder concrete progressively degrades due to the breakage of hydration products such as calcium hydroxide and CSH gel. Above 700°C severe degradation was found to occur in the concrete specimens with partial and full spalling occurrence.

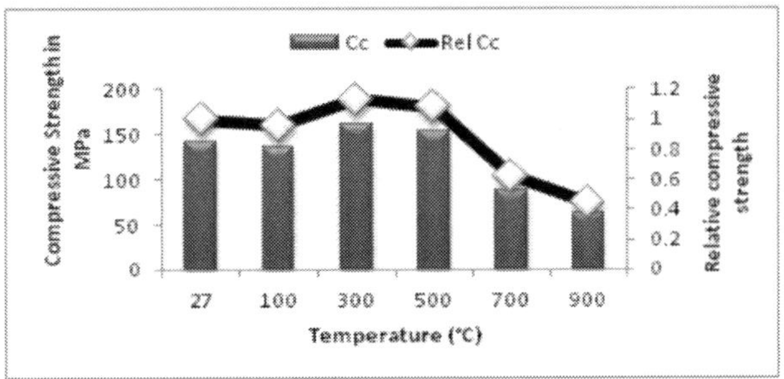

Figure 2 Compressive and relative compressive strength of RPC

3.2. Residual Flexural strength:

Figure 3 illustrates the variation of flexural strength of Reactive powder concrete as a function of temperature. It is evident that the flexural strength decreases linearly with increase in temperature. The flexural strength of Reactive powder concrete decreased sharply up to a temperature of 500°C but decreased slowly from 500°C to 900°C. Almost 50% of the flexural strength was lost at a temperature of 500°C. This severe reduction of flexural strength of Reactive powder concrete at high temperature may be due to their occurrence of micro cracking caused by the built-in pore pressure of heated vapours [16]. Furthermore the sharp decrease in the flexural strength of Reactive powder concrete from 200°C to 400°C may be corresponded to the progressive degradation of calcium hydroxide at around 400°C and disruption of CSH gel around 250°C. Moreover the mass loss associated with the transformation of water available from the hydration products to vapour which induced micro crack formation in the cement matrix.

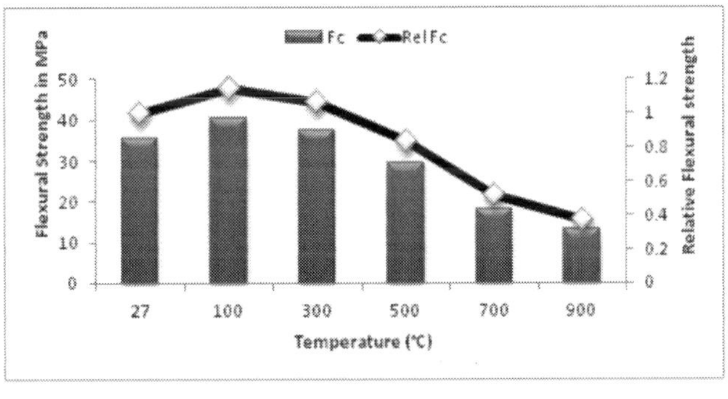

Figure 3 Flexural and relative flexural strength of RPC

3.3. Residual Split tensile strength.

It can be seen from figure 5 that the split tensile strength of Reactive powder concrete was higher than the control samples up to 200°C indicating the better contribution of glass powder and sillimanite towards the tensile behaviour of Reactive powder concrete. This increase in the split tensile strength is attributed to the rehydration of unreacted and unhydrated cementitious particles that gets activated at high temperature [17]. However beyond 200°C, the progressive decrease in the split tensile strength is caused that the Reactive powder concrete migrates from the initial stabilising zone to the strength loss zone where the vapour pressure inside the pores in the concrete intensifies the internal stresses causing a decrease in the residual strength. The test specimen is shown in figure 4.

Figure 4 Split Tensile Strength test

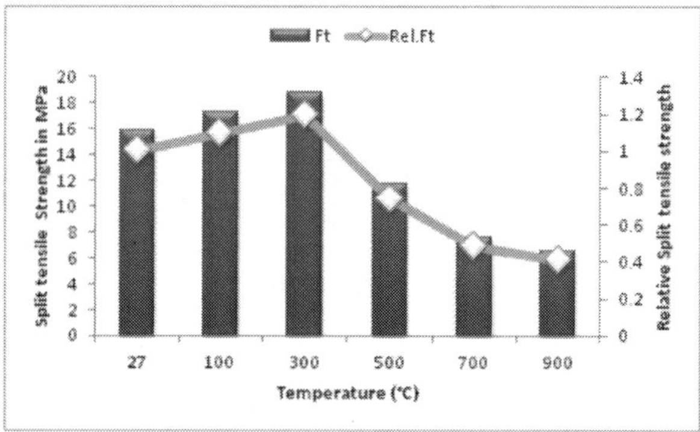

Figure 5 Split tensile and relative spit tensile strength of RPC

3.4 Mercury Intrusion Porosimetry

The porosity of the concrete mixes obtained using water saturation technique shows only the open pores and the inter-granular pores are not determined. The Mercury Intrusion Porosimetry (MIP) technique gives the pore structure and pore sizes present in the concrete. The MIP test was conducted on the concrete containing glass powder and sillimanite at normal temperatures and after exposure to higher temperature. The values were presented in comparison to the control concrete containing no replacements. The values of the pore characteristics of concrete mixes at normal temperature are presented in Table 2.

Table 2. MIP results of the various concrete mixes at normal temperature

MIX ID	Total intrusion volume (ml/g)	Total pore area (m^2/g)	Bulk density (g/cc)	Porosity %
REF	0.0657	7.6254	1.9794	13.3304
MIX1-27^{0}C	0.0604	7.1604	1.8904	12.8364
MIX1-200^{0}C	0.0574	6.5074	1.7944	11.9784
MIX1-400^{0}C	0.0544	5.5034	1.7224	10.1564
MIX1-600^{0}C	0.0514	4.5664	1.6204	9.4784
MIX1-800^{0}C	0.0504	4.2514	1.5424	8.9784

The values clearly show that the porosity was reduced in all the concrete mixes with increasing sillimanite sand substitution. The porosity values and pore area was found to be decreased in the concrete containing sillimanite and glass powder as fine aggregate. It clearly shows that the porosity value was much higher for the control concrete mix. This porosity is a measure of pores that are open to environment and also the pores that are present within the ingredients. The values were found to decrease rapidly with increasing sillimanite sand substitution. The reduced porosity was not only a function of sillimanite but also an effect of glass powder addition that created nucleation sites for hydration reaction. The low particle size of glass powder and relatively finer sillimanite reduced the porosity of concrete. Also the contribution of glass powder towards the secondary hydration reaction and achievement of maximum degree of hydration reduced the porosity of concrete. The CH undergo transformation to CSH increases the solid volume of concrete by minimizing the porosity of concrete. The increasing sillimanite also showed reduced porosity percentage. The porosity values of concrete containing 40% sillimanite sand and 50% sillimanite sand was almost similar with minor variations. Thus it can be inferred that the use of sillimanite sand is a better way for porosity reduction without affecting the strength properties of concrete. The continuous filling of pores by the sillimanite contributing to increase in the hydration of cement can also contribute to the dense microstructure which correspondingly decreases the porosity of concrete.

The results show that the porosity values were reduced for the concrete mixes containing sillimanite sand and this value decreased with increase in sillimanite sand substitution at all temperatures. The slow decomposition of the hydration products at increasing temperature due to the strengthening and reinforcing effect of sillimanite sand at high temperatures contributed to the porosity reduction. The intrusion volume of the concretes containing sillimanite sand was also much reduced indicating minimal pore volume than the control concrete. The chemical transformation of the hydration products CH and CSH was decelerated at high temperatures due to the thermal stability of sillimanite sand which minimized the pore volume at all temperatures.

The results show that the micropores of the concrete reduced significantly due to the sillimanite sand substitution. The macro pores refer to the pores that lie in the range between 107 nm to 1050 nm. The reduction in the macro pores is contributed by the sillimanite sand together. The pore filling nature of sillimanite as well as their ability to densify the microstructure of concrete has also reduced the micro and macro pores of concrete. The control concrete exhibited the highest porosity value when compared to other concrete mixes. The minimum porosity was observed in the micro pores range for the concrete mix containing 50% sillimanite sand as fine aggregate. The particle packing density was increased due to sillimanite sand which resulted in reduced micro pores. The micro pores ranging from diameter 0 to 106 nm was lower in the sillimanite sand substituted concrete mixes indicating the refinement of pore structure of concrete. The concrete is said to be highly permeable when the pores are greater than 100 nm [18] and in the present study the concrete containing sillimanite sand showed minimal pore size at all diameter ranges indicating the impermeable nature of concrete. The reduced pore sizes and intrusion volume shows that the pores were less connected to each other and formed no continuous transport channels. The MIP results clearly showed the improved durability of the produced concrete mixes. The reduced pore sizes and disruption of pores by filling of sillimanite clearly showed the minimal capillary pores through which ingress of external agents may take place. The reduced pore sizes within 100 nm which is the threshold value may be entirely due to sillimanite and the reduction in the macro pores beyond 1050 nm is mainly due to the contribution of sillimanite sand at all temperatures.

3.5 Scanning Electron Microscopic images

Figure 6 shows SEM pictures of the RPC at various temperatures. Because of the physical alterations that happened in the mortar, the specimens became grey approximately 900^{0}C. The glass powder on the surface of the concrete was partially burned, but not totally. The appearance of dark patches in the SEM pictures shows the existence of pores in the concrete. The photos also reveal that the sillimanite's filling action may be responsible for the pore decrease. Furthermore, the flaky layers visible in the concrete photos demonstrate the presence of the CSH gel. The gel-like structure is readily evident at the interface between the glass powder and aggregates, enhancing the cement paste's bonding properties. The enhanced structure of CSH gel induced a considerable decrease in the capillary holes in the concrete by enhancing the interfacial transition zone. The SEM pictures of the mixes from figure 7, likewise revealed well-bonded aggregates with the cement paste, as well as consistent dispersion of the glass powder throughout the cement matrix. The reduced $Ca(OH)_2$ contents acquired from SEM images reflect the above-mentioned findings that the addition of sillimanite contributed to the strength enhancement by converting the $Ca(OH)_2$ to CSH gels. Normal concrete had some CSH gels that were not fully formed and were thinner than the glass powder and sillimanite substitution concrete mixtures.

The combined impact of sillimanite and glass powder converted the CH crystals into CSH gels that encircled the tiny aggregates, resulting in a denser and more compact micro-structure. The existence of deformed CSH gels was clearly visible in SEM images taken at 900°C, indicating a larger development of CSH gels due to crystalline structure degradation in the $Ca(OH)_2$ and CSH crystals. The increased porosity of the thermally treated reactive powder concrete mixes also reveals disturbed CSH gels and deformed CH crystals, resulting in a decrease in strength values.

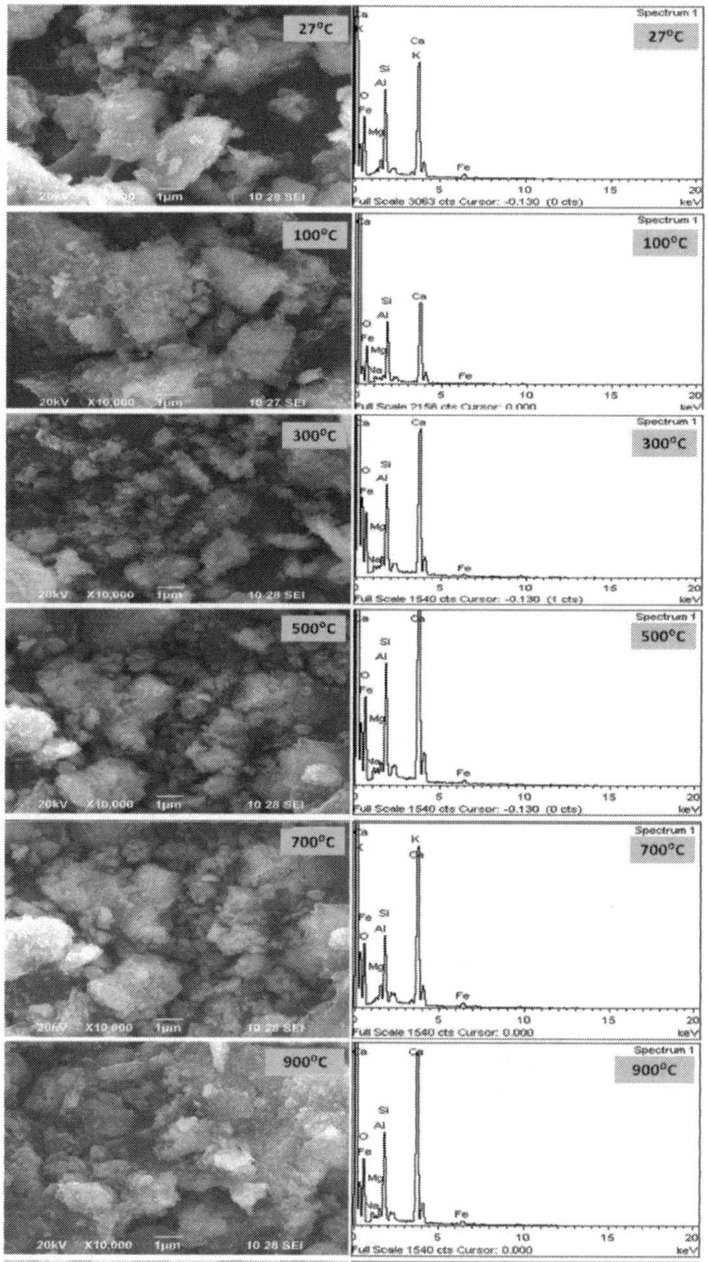

Figure 6 SEM-EDAX results of control RPC at various temperature

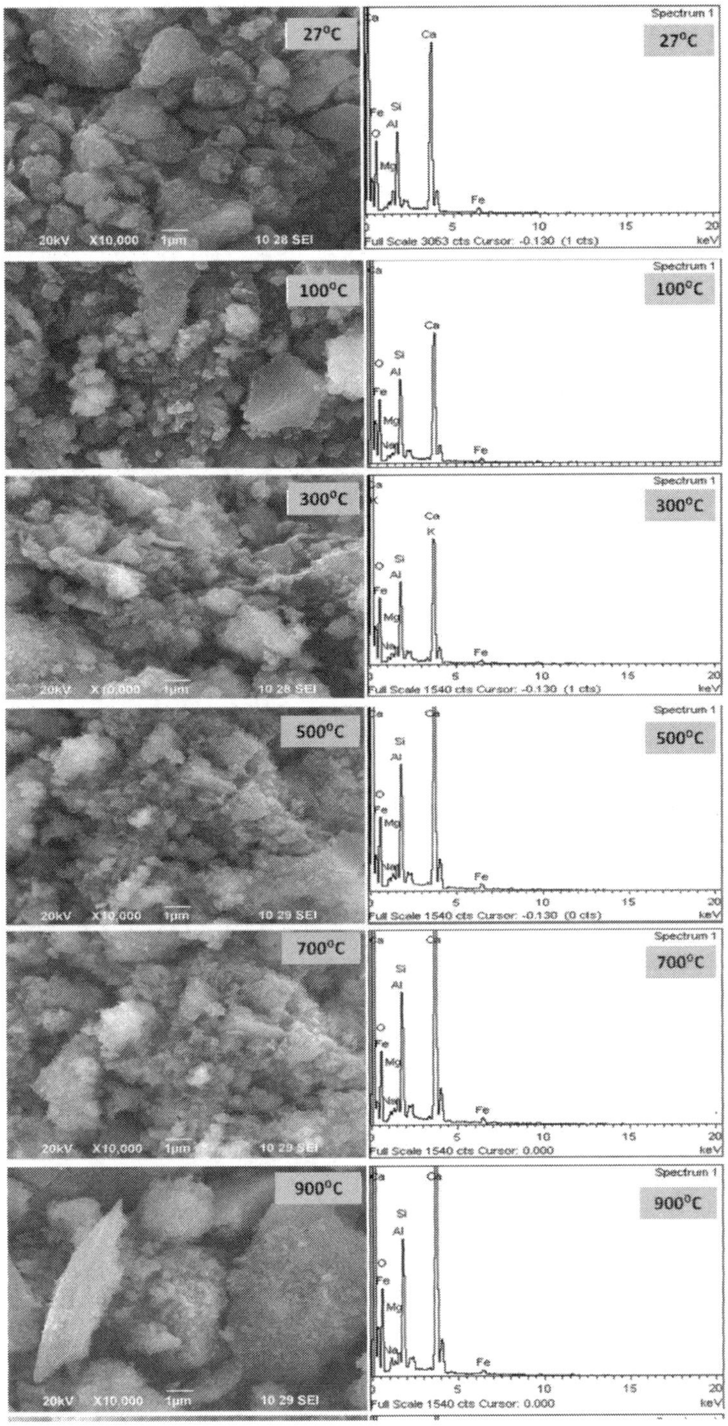

Figure 7 SEM-EDAX results of mix1 at various temperature

4. Conclusion

Based on the experimental results of Reactive powder concrete using glass powder and sillimanite as fine aggregate after exposure to high temperature the following conclusions can be summarized:

1. The compressive strength increase initially with increase in temperature up to 300°C. However beyond 300°C the compressive strength of Reactive powder concrete was lowered.
2. The flexural strength loss was observed for all the samples after exposure to elevated temperature and when the temperature was about 800°C the Reactive powder concrete lost almost all of its flexural strength.
3. The Reactive powder concrete after exposure to elevated temperatures resulted in the reduction of split tensile strength however the reduction occur only after the temperature was raised above 400°C.
4. The impact of sillimanite with glass powder and RPC components on concrete porosity reduction is substantial because the densification of the microstructure and compactness of concrete has been greatly increased.
5. The morphology of the concrete formed with sillimanite, glass powder, and quartz sand improved with rising temperature compared to the standard mortar. The inclusion of sillimanite resulted in an impermeable cement matrix that tightly bound the aggregate, resulting in dense interfacial transition zones. The size and volume of the pores in the generated RPC mixes decreased with increasing temperature, showing the matrix densification effect induced by sillimanite addition.

Generally the dense microstructure of reactive powder concrete caused explosive spalling after exposure to high temperature. However in the present study the addition of sillimanite and glass powder as fine aggregate greatly improved the high temperature resistance of reactive powder concrete mitigating the bursting and explosive spalling caused by silica fume and fine materials at high temperature.

5. References

[1] Messina , Ferone , Colangelo, Roviello, Cioffi, "Alkali activated waste fly ash as sustainable composite : Influence of curing and pozzolanic admixtures on the early-age physic-mechanical properties and residual strength after exposure at elevated temperature", Composites Part B,vol:132,pp:161-169,2018.
[2] Chin - Tsuang Liu , Jong – Shin Huang , "Fire performance of highly flowable reactive powderconcrete", Construction and Building Materials, vol:23,pp:2072-2079,2009.
[3] Nima Farzadina , Abang Abdullah Abang Ali , Ramazan Demirboga, "Characterization of high strength mortars with nano alumina at elevated temperatures",Cement and Concrete Research ,vol:54,pp:43-54,2013.
[4] Raje Gowda[a], H Narendra[a], Dinesh Rangappa[b*], R Prabhakar[a*], "Effect of nano –alumina on workability , compressive strength and residual strength at elevated temperature of cement mortar", Materials Today , vol:4,pp:12152-12156,2017.
[5] Msheer Hasan Ali, Youkhanna Zayia Dinkha ,James H. Haido, "Mechanical properties and spalling at elevated temperature of high performance concrete made with reactive and waste inert powders", Engineering Science and technology an international Journal ,2016.
[6] Mehmet Canbaz , "The effect of high temperature on reactive powder concrete", Construction and Building Materials, vol:70,pp:508-513,2014.
[7] Muhammad Abid[a,b] , Xiaomeng Hou[a,b], Wenzhong Zheng [a,b,*], Raja Rizwan Hussain, "High temperature and residual properties of reactive powder concrete-A review",vol:147,pp:339-351,2017.
[8] A.Cwirzen[*] ,The Effect of the heat – treatment regime on the properties of reactive powder concrete", Advances in cement research , vol:19,pp:25-33,2007.
[9] Peng Yanzhou[a,b,c,*], Zhang Jun[b,c], Liu Jiuyan[b,c], Ke Jin[b,c], Wang Fazhou[a], "Properties and microstructures of reactive powder concrete having a high content of phosphorous slag powder and silica flume", Construction and Building Materials ,vol:101,pp:482-487,2015.
[10] Yin – Wen chan[*] , Shu-Hsien Chu , " Effect of silica fume on steel fiber bond characteristics in reactive powder concrete", cement and concrete Research,vol:34,pp:1167-1172, 2004.
[11] Muhammad Abid[a,b] , Xiaomeng Hou[a,b], Wenzhong Zheng [a,b,*], Ghulam Quadir Waqar , "Mechanical properties of steel fiber – reinforced reactive powder concrete at high temperature and after cooling", Procedia Engineering , vol:210,pp:597-604,2017.

[12] Yulong Wang , xiangcheng Li*, Pingan chen , Boquan zhu* , " Matrix microstructure optimization of high alumnia-special castables and its effect on High temperature properties", Ceramics International,2017.
[13] Haiyan Li,and Gang Liu , Tensile Properties of Hybrid Fiber- reinforced Reactive powder concrete after exposure to Elevated temperatures", International journal of concrete structures and Materials , 2016.
[14] Keke Huang , Yuambing Li, Shujing Li, Li Wang , Shihan Wang , "Effects of using zicora sand as partial substitute of fully refractory powdered concrete at elevated temperatures",Ceramics international.
[15] Muhammad Abid[a,b] , Xiaomeng Hou[a,b], Wenzhong Zheng [a,b,*] , Raja rizwan Hussain[c] , " Hiogh temperature and residual properties of reactive powder concrete – A review", Construction and Building materials,vol:147,pp:339-351,2017.
[16] Yang Ju, kaipei Tian, HongbinLiu , Hans- Wolf Reinhardt, Li Wang , "Experimental investigation of the effect of silica fume on the thermal Spalling of reactive powder Concrete", Construction and Building materials,vol: 155,pp:571-583,2017.
[17] Sharmarke abdi omer, Ramazan Demirboga , waleed H. Khushefati[a], " Relationship between compressive strength and Upv of GGBFS based geopolymer mortars exposed to elevated temperatures", Construction and Building materials,vol:94,pp:189-195,2015.
[18] Josef Novak, Alena Kohoutkova, "Mechanical properties of concrete composites subject to elevated temperature", vol:95,pp:66-76,2018.

Journal of Physics: Conference Series **2332** (2022) 012015 doi:10.1088/1742-6596/2332/1/012015

Characterization of Biochar Bricks to be used as a Construction Material

B. Nithyalakshmi[1], Soundarya. N[2], Praveen S[3]
[1] Assistant Professor, Coimbatore Institute of Technology, Coimbatore
[2] Associate Professor, CMR Institute of Technology, Bengaluru
[3] Assistant Professor, CMR Institute of Technology, Bengaluru

soundarya.n@cmrit.ac.in

Abstract. The experimental work on characterization of biochar bricks as a construction material is summarized in this manuscript. This is an important topic to research due to the dearth of knowledge about biochar construction materials and their potential to reduce global warming. Biochar is a charcoal-like substance formed from organic matter decomposition in the absence of oxygen at high temperatures. The use of bio char in a range of applications will help in cutting down CO_2 emissions, recycling of organic and plastic waste, and in creation of more jobs. Current applications of bio-char is limited to being used as Soil additives to promote water absorption, plaster to absorb humidity, and energy alternatives to replace fossil fuels. The present study attempts to characterize bio-char bricks for usage in construction. Two types of bio-char bricks were cast and compared to normal concrete bricks. The first set of bricks had 70% biochar and 30% cement, whereas the second set contained 45 percent biochar, 45 percent plastic, and 10% cement. The bricks were characterized using five tests: compressive strength, water absorption, hardness, flammability, and insulation value.

1. Introduction

Pyrolysis is a process of burning substances at a higher temperature in the absence of oxygen. This process inhibits organic matter from interacting with oxygen as it decomposes, preventing carbon dioxide and carbon monoxide generation. As a result, a solid carbon structure forms, trapping the carbon from organic matter in solid form, preventing breakdown and the generation of methane, carbon monoxide, or carbon dioxide. This reduces the amount of carbon dioxide that enters the atmosphere [1]. Biochar is one such product. Biochar is a charcoal-like substance made from decomposition of organic materials in the absence of oxygen at high temperatures. Biochar has a very low thermal conductivity as it is mainly made of carbon [2]. Biochar has the ability to absorb five times its own weight in water. This is owing to the porous nature of biochar's structure, which varies depending on the temperature at which it is produced and the type of organic material utilized as a source material. Currently biochar is being used to enhance the fertility of soil, controlling methane emissions in dairy farms and as water filters [2, 3]. The use of biochar in building materials is a new line of research that is being pursued in the recent times. Successful use of biochar in the construction industry can greatly help the environment in two ways. Biochar serves two purposes: first, it conserves virgin resources and second, it traps greenhouse gases like methane and carbon dioxide in solid carbon form [4, 5]. The current work focuses on using biochar to make masonry bricks for construction purposes. Concrete bricks were used as the reference material. Two types of experimental bricks were casted, one with a combination of biochar and cement and other with the combination of biochar, plastic and cement. The percentages added were determined through pilot experiments, and these two were chosen for further research based on the finish of the bricks and the compressive strength obtained. All bricks were assumed to be homogenously mixed with an even surface. It was assumed that particle size of any material used had no effect on the mixing of the bricks.

2. Materials and Methodology

This section explains the materials used and the methodology with which the research was carried out.

2.1. Materials

Cement, biochar, fine aggregates, course aggregates and plastics. The physical properties of the materials are shown in the table 1. The control mix casted was of M20 concrete and the proportion is shown in table 2. The experimental bricks casted are detailed in table 3. The bricks were of the size 200mm x 100mm x 60mm.

Table 1. Physical Properties

Material	Specific Gravity
Cement	3.15
Fine aggregate	2.65
Coarse aggregate	2.8
Biochar	1.6
Plastic ((HDPE) Type 2 Recycled Plastic)	0.95

Table 2. Control Mix

Type of Brick	Cement	Fine Aggregates	Coarse Aggregates	W/C Ratio
CC	1	1.5	3	0.55

Table 3. Experimental Bricks Casted

Type of Brick	Cement %	Biochar %	Plastic %
BC 1	30	70	-
BC2	10	45	45

The HDPE, Type 2 Plastic were collected from waste and cut into pieces. The pieces were submerged in liquid nitrogen for 5 minutes and then was ground to powder form in a blender. This is to ensure it mixes well with the biochar.

2.2. Methodology

Five tests were carried out for the characterization.

2.2.1. Compressive Strength Test

Knowing the compressive strength of bricks is critical for determining their suitability for construction. With the use of a compression testing machine, compressive strength tests on bricks were performed to assess the load carrying capacity of bricks under compression.

2.2.2 Insulation value Test

The R-Value, or insulation value, is a measurement of a material's ability to resist heat flow. This test compares the performance of biochar bricks to that of regular bricks. For this a reference material was used in this current work to measure the accuracy of the test. The reference material used was oak wood brick. The temperature difference between the cold and hot surface of this reference material was calculated and was compared with the existing standards to measure the accuracy. The tests were carried out on the experimental bricks once the accuracy was validated.

2.2.3. Water absorption Test

Water absorption is vital to investigate not just to determine the brick's endurance in humid or high-rainfall environments, but also to determine the material's permeability. The bricks were immersed in potable water for 24 hours at room temperature. The weight was recorded, before and after the immersion.

2.2.4. Flammability

For building materials that use plastics flammability is an important value. It is critical to examine the flammability characteristics of the material in order to understand essential characteristics such as the lower flammability limit and higher flammability limit in order to reduce the danger of fire or explosion. The experimental set up consisted of a metal stand to place the test brick. The torch of the burner was fastened into place with the nozzle one inch away and orthogonal to the brick face. This was done with the help of a ring stand and a three-pronged clamp. A torch with a three-foot hose was attached and used to position the propane tank a safe distance away from the flame. The ignition and extinguish time tests were the first step of the experiment. By depressing the trigger and employing the lock mechanism to avoid direct contact with the torch during the experiment, the torch was turned on until the brick face was confirmed to be ignited. To establish an ignition time, the burn process was timed. Because it's impossible to see if a brick with a flame on it is sustaining its own flame in this test, the differentiating feature of an ignited face was material disintegrating from the flame. The torch was switched off once the brick was ignited, and the time it took for the brick to extinguish was recorded. The time when the brick ceased blazing red with embers was considered to be the extinguish time.

2.2.5. Hardness

The ability of a substance to resist erosion and surface deformation is determined by its hardness. A Shore D durometer was used to determine the bricks' hardness. The durometer's tip was pressed into the sample five times on the top and five times on the bottom, evenly spaced around the brick, to obtain an average hardness. The durometer was used to take a measurement by pressing the tip of the durometer into the sample while maintaining a uniform and even pressure on the durometer with one's hand. The durometer tip was placed perpendicular to the sample, and measurements were collected from various parts of the sample brick.

3. Results and discussions

Concrete bricks of M20 strength was used to compare the characteristics of the biochar bricks. Two bricks with biochar was casted along with different materials, BC1 – 70% Biochar and 30% Cement and BC 2 – 45% Biochar, 45% Plastic and 10% Cement.

3.1 Compressive Strength test

On the 7th and 28th days, concrete bricks, C1 had a strength of 10Mpa and 18Mpa, respectively, BC1 showed a strength of 12Mpa and 10Mpa, respectively, whereas BC2 showed a strength of 25Mpa on the 28th day as shown in Table 4 and 5.

Table 4. Compression Testing Result of Concrete Sample

Name	Type of brick	Composition	Days	Compressive Strength (Mpa)
C1	Concrete	1: 1.5: 3 0.55 W/C	Day 07	10
			Day 28	21

It can be noted that there is a slight decrease in the compressive strength in BC 1 from 7th day to 28th day. Compressive strength directly relates to the structural integrity of the bricks. The lower compressive strength can be attributed to the softness of the biochar material on comparison with fine aggregates. Also it is possible that the reduced compressive strength of BC 1 might be due to the premature cracking between the 7th day to 28th day. The particle size distribution has been done and reported in the figure 1. BC2 has given a higher compressive strength of 25Mpa. This can be attributed to the homogeneity of the Biochar brick sample due to the addition of plastics and reduction in the preterm cracking [4]. Also BC2 shows higher strength because, the bricks didn't give a brittle failure like normal bricks. They kept getting deformed without any visible cracks, they kept deforming and compressed as the loads increased [5].

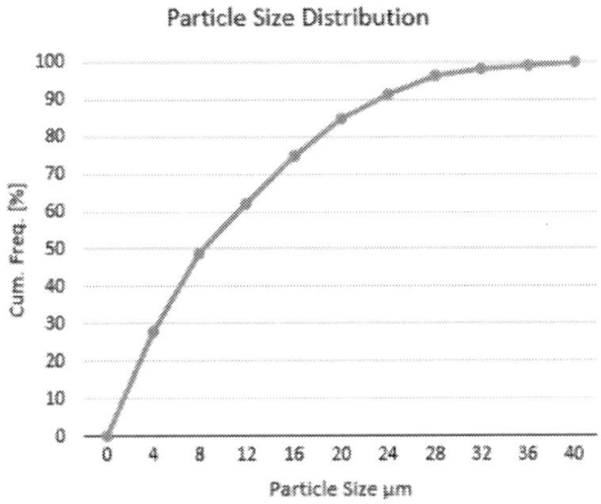

Figure 1. Particle Size distribution of Bio-char

Table 5. Compression Testing Result of Biochar bricks

Name	Type of brick	BC	Plastic	Cement	Days	Compressive Strength (Mpa)
BC 1	70% BC 30% Cement	70	0	30	Day 07 Day 28	12 10
BC 2	45% BC, 45% Plastic and Cement 10%	45	45	10	Day 28	25

3.2 Insulation Value

Table 6. Insulation value.

Sample Name	Days	Insulation Value (R- Value)
C1	Day 7	0.6
	Day 28	1.2
BC1	Day 7	0.8
	Day 28	0.7
BC2	Day 28	0.78

The insulation value is denoted as R- Value. Both BC1 and BC2 had lesser insulation values on comparing with C1 as shown in table 6. BC1 and BC2 also show an R- Value of 0.7 and 0.78 and that is because of the presence of carbon in the biochar bricks. In both BC1 and BC2 it can be noted that there is a drop in the insulation value from day 7 to day 28. This can be attributed to the chemical reaction between the cement particles and the bio- char particles [6].

3.3 Water Absorption

In third-world and developing countries, water absorption is critical for determining if the brick is feasible under humid or high-rainfall circumstances.

Table 7. Water absorption percentage.

Type of brick	Water Absorption %
C1	6 5
BC1	44 40
BC2	20

It was observed that the biochar bricks have a greater water absorption on compared with the concrete bricks as shown in table 7. This is because of the porous nature of the biochar bricks. More pores means more surface area for water to get absorbed [7].

3.4 Flammability

Table 8. Flammability.

Type of brick		Burn Time (Secs)	Extinguish Time (Secs)
BC1	Day 7	55	55
	Day 28	140	75
BC2	Day 28	75	50

The flammability parameters of materials are critical in determining their structural safety. The ignition, extinguish, and separation times of the bricks casted were noted down and reported in table 8. A C-clamp was utilized to stage the brick, in the apparatus. The results on day 9 were inconclusive because the brick fell before the separation test was completed due to the C- Clamp. No smoke was observed during the tests. For BC1 on 7th day the burn and extinguish times were 55 and 55. On 28th day it was 140 and 75. The difference in the burn and extinguish time is large on day 28 may be due to the curing of bricks. The large difference in burn time on 7th and 28th day is because of various factors like curing of the cement and also the reactivity of the cement with biochar particles [8]. Same trend can also be noted in BC2 on the 28th day and it is hence, safe to say that curing has a greater effect on the biochar bricks because the structure of the bricks become sturdier and can resist decomposition better [9].

3.5 Hardness

Table 9. Hardness Value.

Type of brick	Days	Top	Bottom
	Day 7	65	63
C1	Day 28	80	83
	Day 7	40	60
BC1	Day 28	42	55
BC2	Day 28	-	-

On comparing the biochar bricks with the concrete bricks it can be noted that the biochar bricks BC1 have a lowered value of hardness as shown in table 9. It can be attributed to the porous nature of BC1 [10, 11]. It can also be noted that concrete bricks have harder materials in its composition on comparing with the softer biochar bricks, due to the material composition also the hardness can be affected. BC2 was very soft and hence a value wasn't obtained [12, 13].

4. Conclusions

Biochar bricks were compared to regular concrete bricks in an attempt to characterise them. The first batch of bricks had 70% biochar and 30% cement, while the second batch contained 45 % biochar, 45 % plastic, and 10% cement. Compressive strength, water absorption, hardness, flammability, and insulating value were all tested on the bricks. BC1 rather than BC2 can be considered a potential building material based on the results. BC2 was unable to do as well in the hardness test as it did in the compression test[7]. So, with a little more research and modification to make BC1 perform better in water absorption and flammability tests, the material might be a great alternative to traditional concrete bricks.

References

[1] 'Biochar as Building Material', Ithaka Institute - Building Material, Ithaka Institute, www.ithakainstitut.org/en/ct/97.

[2] 'Basic Information about Landfill Gas', EPA, Environmental Protection Agency, 18 Dec. 2019, www.epa.gov/lmop/basic-information-about-landfill-gas

[3] Wang, T. "U.S. and World Cement Production 2018." Statista, 9 Aug. 2019, www.statista.com/statistics/219343/cement-production-worldwide/

[4] Potential Uses of Biochar." Access Science, McGraw-Hill Education, 1 Jan. 1970, www.accessscience.com/content/potential-uses-of-biochar/BR111816

[5] Ahmad Z, Mosa A, Zhan L, Gao B (2021) Biochar modulates mineral nitrogen dynamics in soil and terrestrial ecosystems: a critical review. Chemosphere 278:130378

[6] Ivory Coast Is Using Plastic Waste to Build Schools." BBC News, BBC, 23 Jan. 2020,www.bbc.com/news/av/world-africa-51213133/ivory-coast-is-using-plastic-waste-to-build-schools.

[7] Biochar Bright Spots: Biochar Bricks ." Finger Lakes Biochar, Finger Lakes Biochar, 10 Jan. 2019, fingerlakesbiochar.com/biochar-bright-spots-biochar-bricks/.

[8] NRMCA. "CIP 35 - Testing Compressive Strength of Concrete." National Ready Mixed Concrete Association, 2003, www.nrmca.org/aboutconcrete/cips/35p.pdf

[9] Paglierani, Gary. Hardness Testing of Materials. CSU - Fresno, lennon.csufresno.edu/mhk03/hardness1.htm.

[10] Praveen, S., Jegan, J., Bhagavathi Pushpa, T. et al. Biochar for removal of dyes in contaminated water: an overview. Biochar 4, 10 (2022). https://doi.org/10.1007/s42773-022-00131-8

[11] Gokulan R, Praveen S, Rajamurugadoss and Jegan J, 2022, Water—Conventional and novel treatment methods, Solar-Driven Water Treatment Re-engineering and Accelerating Nature's Water Cycle pp. 37-66 https://doi.org/10.1016/B978-0-323-90991-4.00003-7

[12] Saravanan, P, Josephraj, J, Thillainayagam, B, P, Environ. Nanotechnol. Monit. Manag. 2021, 16, 100560 https://doi.org/10.1016/j.enmm.2021.100560

[13] Kezhen Qian, Ajay Kumar, Hailin Zhang, Danielle Bellmer, Raymond Huhnke, 'Kezhen Qian, Ajay Kumar, Recent advances in utilization of biochar', Renewable and Sustainable Energy Reviews, 2015, vol 42, 1055- 1064.

Bandgap Variations and Photodegradation Efficiency on Inclusion of SiO_2 in TiO_2-rGO Nanocomposite for High Performance Photocatalysis

S Preetha [1,2,3] and I C Lekshmi* [1]

[1]Department of Chemistry, CMR Institute of Technology, Bengaluru, India
[2]VTU-RC affiliated to Visvesvaraya Technological University, Belagavi, Karnataka, India
[3]Department of Chemistry, Cambridge Institute of Technology, Bengaluru, India

*Corresponding Author:

chaitanya.l@cmrit.ac.in, ic.lekshmi@gmail.com.

Abstract. SiO_2 is an excellent support material due to its flexible pore size and can be used as an adsorbent for the removal of toxic materials from water. Nanocrystalline TiO_2-rGO is previously reported to show complete photodegradation for azo dyes. In pursuit of enhancing the adsorption capability of the system, we have incorporated SiO_2 in TiO_2-rGO nanocomposite which has drastically reduced the crystallinity of the matrix. Analysis of the bandgap variations is pursued here, which is crucial for enhanced light absorption and high performance photocatalysis.

Keywords: SiO_2, TiO_2-rGO, Amorphous, Bandgap variations, photodegradation

1. Introduction

One of the major disadvantages of using TiO_2 as a photocatalyst is the turbidity generated in their aqueous solutions through dispersion that inhibits the UV light transmission. The slurry form of TiO_2 also poses difficulties in separating the catalyst when the photodegradation reactions are completed [1]. The addition of mesoporous silica to TiO_2 is reported to help in catalyst separation besides advantageously modifying the surface hydroxyls, surface area, and thermal stability [2, 3]. SiO_2 is also an excellent support material due to its flexible pore size and can be used as an adsorbent for the removal of toxic materials [2, 4]. However, on adding silica to titania matrix, the particle size was reported to be reduced and the specific area enhanced, thus increasing the photocatalytic activity [4]. Carbon nanotubes functionalized with TiO_2/SiO_2 are also promising for photocatalytic decomposition of different organic contaminants in water [5].

When rGO nanosheets are combined with TiO_2 or SiO_2 NCs, their interfacial contact and the creation of conductive routes that lead to suppression of charge recombination, have hybrid features of both nanomaterials and graphene [6]. rGO acts as an efficient cover and protection for the catalyst species, increasing their surface area and improving the charge carrier separation and transport [7]. High photoactivity was reported for 47% rGO–TiO_2 nanocomposite (TG4) in our previous study [8]. We are investigating here the TiO_2–rGO–SiO_2 (TGS) nanocomposites for improving the performance of the photocatalyst. The addition of SiO_2 into the TG4 matrix is studied in terms of structure, morphology and photodegradation behaviour as different amounts of tetraethyl orthosilicate (TEOS) is added. The synthesis protocol used ascorbic acid as an effective and environment-friendly reducing agent to obtain TG4 and TGS samples. Using Alizarin Yellow GG (AYGG) azo dye as a model organic pollutant, we also demonstrate the good light energy absorption and degradation for TGS catalytic system.

2. Experimental Methods

2.1 Nanocomposite Synthesis

GO was synthesised employing the modified Hummer's method and was chemically reduced to rGO. TG4 nanocomposites were further prepared using ascorbic acid as green reducing agent. The specific details of the above could be found in our previous study [8]. For TGS nanocomposites, a fixed amount of GO was initially added to 100 ml of ethanol and mixed for 2 h. Variable volumes of TEOS solutions (1 ml, 2 ml, 3 ml and 4 ml) were added separately to four distinct sets of GO solutions with comparable volumes and specified amounts (mole) of TTIP based on weight % ratio at 40^0C. Ascorbic acid (0.01 M) was added to each as a reducing agent and stirred for 18 h. The gel obtained was aged at RT, filtered, rinsed in deionized water and dried. Four distinct nanocomposites TGS1 (50%TiO_2, 44%rGO, 6%SiO_2), TGS2 (47%TiO_2, 42%rGO, 11%SiO_2), TGS3 (44.5%TiO_2, 39.5%rGO, 16%SiO_2) and TGS4 (42%TiO_2, 38%rGO, 20%SiO_2) were then calcined at 400°C.

2.2 Characterization

The prepared samples were characterized for structural, morphological, electronic and photodegradation details using X-ray diffraction (XRD), Transmission electron microscopy (TEM), X-ray photoelectron spectroscopy (XPS), UV-Diffuse Reflectance Spectra (UV-DRS) and direct sunlight photodegradation studies.

The XRD patterns of the samples were analysed using a table top powder X-ray diffractometer, Bruker D2 Phaser, using CuKα radiation and a Ni filter in the 2θ scan range of 10-80°. The produced nanocomposites and NCs were imaged using TEM. 50 mg of each sample was dissolved in 25 ml of IPA and ultrasonically agitated for 20 min to achieve full dispersion for a standard study. Then, under

Journal of Physics: Conference Series **2332** (2022) 012016 doi:10.1088/1742-6596/2332/1/012016

a table light, the dispersed solution was drop-casted on 300 mesh copper TEM grids and the coated grids were kept in the holder of the microscope. The absorption spectra of the samples were analysed and their bandgap calculated using UV-Vis-DRS data collected using a Lambda 365 Perkin-Elmer UV-Visible spectrophotometer and 50 mm integrating sphere at room temperature. BaSO4 was used as a reference. UV-DRS is used to analyse the powdered solid sample once the wavelength range, mode (absorbance/reflectance), and scan control have been configured. The elemental composition of the samples as well as the chemical and electronic states of the constituent elements were identified using XPS. A Thermo Scientific (UK) spectrometer with an Al K-alpha X-ray source and a C1s correction at 284.6 eV for calibration was used for data capture.

For photodegradation experiments, a 100-ppm AYGG dye stock solution was prepared and diluted to 10 ppm for photoconversion experiments with TGS4 nanocomposite as catalyst with a loading of 100 mg per 200 ml dye solution. The initial dark reactions were performed to establish process equilibrium utilising a fully wood-built light protection device. The wooden box contained a magnetic stirrer and a reaction setup, and the system was swirled for 30 minutes at the onset. An additional 30 min was allowed without stirring to reach adsorption-desorption equilibrium also conducted in dark. The reaction setup was then exposed to 1,18,000 to 1,32,000 lux sunlight intensity (bright sunny day) for photocatalytic degradation testing that lasted for 150 min. A duplicate of the reaction system was also set-up and was left in dark inside the box for the same time period of 150 min to complete the adsorption reaction.

At 15-min intervals, 3 ml aliquots from both sample sets were withdrawn using a SEOH syringe and examined for changes in AYGG dye absorption spectra excited in the wavelength region 180-800 nm. The absorption spectra were checked for distinctive band maximum at 354 nm characteristic of n-π^* electronic transition associated with the azo group in the dye molecule. The tests were conducted using the dye solution in presence of the catalyst at its natural pH of 6.8. The degradation of AYGG dye was calculated using the equation,

$$\%\ \text{Degradation} = \frac{C_0 - C_t}{C_0} \times 100$$

where C_0 and C_t are the concentration of the dye solution at time t = 0 and t min, respectively.

3. Results and discussion

Pure anatase phase of TG4 was previously confirmed by the corresponding XRD pattern in Fig. 1(a) [8]. In TGS nanocomposites, the intensity of TiO_2 peaks were drastically reduced and changed to single peak that was amorphous in nature, as SiO_2 concentration was increased. The single peak at 25.3° also overlapped with those of rGO (002) and TiO_2(101) in all the TGS nanocomposites [8], and no distinct peaks of rGO and SiO_2 were seen. Additionally, the (100) diffraction peak of rGO at 43° vanished in the TGS nanocomposites [9]. Compared to TG4, the peak maxima shifted to higher 2θ values on inclusion of SiO_2 indicating changes in crystal features such as lattice parameters, crystallite sizes and d-spacing. The presence of amorphous SiO_2 in contact with TiO_2 can effectively inhibit the growth of the latter nanocrystals (NCs) through formation of Ti-O-Si bonds, as could be seen from the XRD patterns. Due to the low crystallinity of TGS systems, it was difficult to calculate the crystallite sizes using Debye Scherrer's equation. Extremely tiny particles of TiO_2 and SiO_2 embedded in rGO matrix could be seen from the TEM image of TGS4 given in Fig. 1(b).

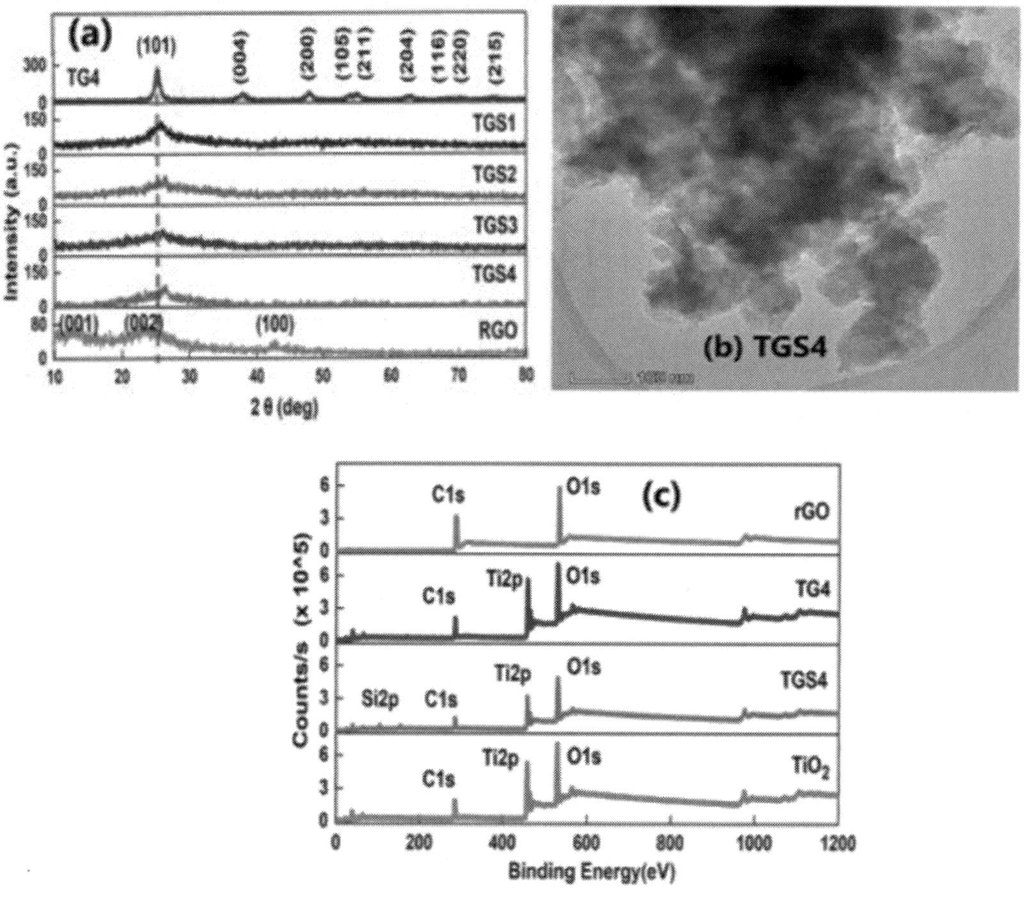

Figure 1 (a) XRD patterns of rGO, TiO_2 NCs, TG4 and TGS nanocomposites; (b) TEM image of TGS4 nanocomposite; (c) Survey XPS spectra of samples

The inclusion of C, O, Si, and Ti elements in TGS4 composites was confirmed by EDS analysis (not shown here). The XPS survey spectra of the compounds examined here are shown in Fig. 1(c). On addition of SiO_2, TGS4 spectra exhibited a significant drop in the intensity of the C1s, O1s, and Ti2p peaks related to the reduction in the percentage content of rGO and TiO_2 components in the nanocomposite. Si2p peak could be located at 102.3 eV confirming the presence of silicon. More detailed analysis of core-level elemental spectra is underway to determine the local structure changes of TGS nanocomposite.

The absorption spectra recorded for bare TiO_2 NCs, TG4 and TGS nanocomposites at RT can be found in Fig. 2(a). The absorption edge in TGS nanocomposites was sharper and the absorption intensity was higher compared to TiO_2 and TG4. The absorption intensity was also high for TG4 compared to TiO_2 NCs due to incorporation of high rGO content, though less than most TGS samples. The bandgap value of the samples was shown alongside the spectra in each case, and was determined applying the Kubelka–Munk function to DRS spectra and plotted as a variation of absorption energy. Accordingly, the Eg values for bare TiO_2, TG4, TGS1, TGS2, TGS3 and TGS4 nanocomposites were found to be 2.96 eV, 3.08 eV, 3.22 eV, 3.3 eV, 3.27 eV and 3.32 eV respectively. The addition of more insulating SiO_2 decreased the size of TiO_2 NCs in TGS nanocomposites resulting in an increase in their bandgap compared to TG4.

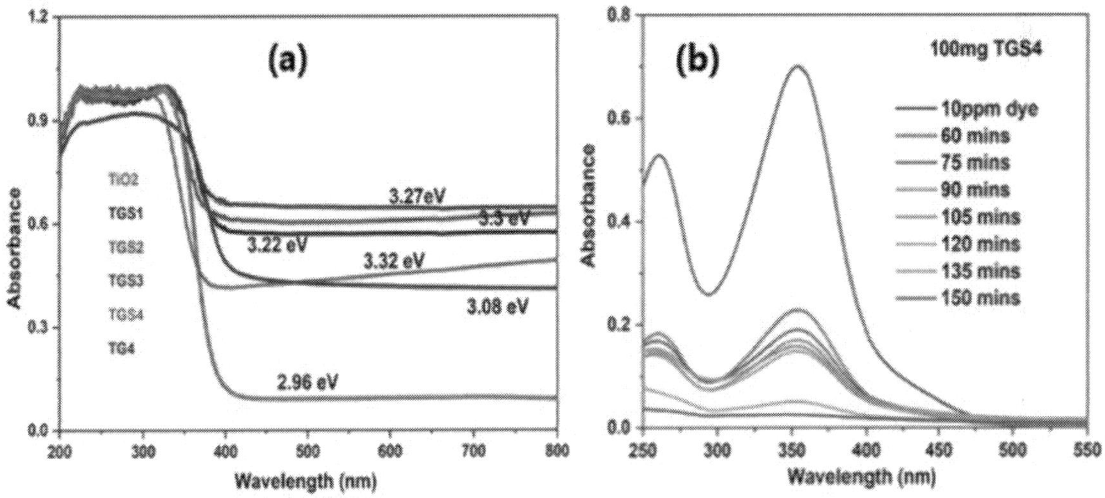

Figure 2 (a) UV-DRS absorbance spectra of TiO_2 NCs, TG4, and TGS nanocomposites; (b) Photodegradation of 10 ppm AYGG dye at different irradiation time in presence of TGS4 catalyst for a loading of 100 mg/ 200 ml dye solution as indicated by reduction in UV-Vis absorbance peak

The photodegradation of AYGG azo dye was examined to verify and compare the effectiveness of TGS nanocomposite as a photocatalyst compared to TG4. UV-Vis absorbance peak of 10 ppm AYGG dye at different irradiation time in presence of TGS4 catalyst (best among TGS nanocomposites) is shown in Fig. 2(b). A drastic decrease in absorption maxima was seen in first 60 min followed by a gradual decrease up to 96% in 150 min. This was close to 100% degradation reported for the same dye using TG4 photocatalyst used in our previous study, and conducted under similar experimental conditions. The result was also much higher than 80% degradation shown by TiO_2 NCs [8]. The adsorption of TGS4 catalyst at the end of 150 min in dark was 67.27% (not shown here). The inclusion of SiO_2 did not comprise much on the total efficiency of TG4 photocatalyst under direct sunlight, though it helped in much better catalyst separation post reaction completion.

4. Conclusion

The rate of AYGG degradation by TiO_2 was accelerated in the presence of rGO. To have a smart catalyst that could be easily separated post degradation reaction, we incorporated a small amount of SiO_2 (upto 20%) that changed the nanocrystallinity of original TG4 nanocomposite. Though a raise in bandgap to deeper UV zone was observed on SiO_2 inclusion, the intensity of light absorption was not comprised, and in fact it was enhanced. This needs a detailed investigation, which we are currently undertaking. Qualitatively, rGO is effective in collecting photoexcited electrons and preventing electron-hole recombination. This could effectively speed up charge transport between TiO_2 or SiO_2 and rGO following photoexcitation, retaining the photocatalytic efficiency. The observation was supported by the strong SiO_2 dispersion and interfacial coupling between TiO_2 and SiO_2 with rGO in the sheets as seen from their TEM images. While rGO lowers the rate of electron-hole pair recombination, SiO_2 enhances the surface hydroxylation and surface area of TiO_2 NCs, besides providing good separation facility post reaction completion. At the moment we are unsure of the impact of low crystallinity of the photocatalyst on the catalyst stability and recyclability. We are undertaking detailed studies for a thorough analysis of the TGS4 nanocomposite catalyst system.

Acknowledgement

Dr. Chaitanya Lekshmi Indira is grateful to Department of Science and Technology for Nanomission (Grant No. SR/NM/NS-1161/2013) and Young Scientist Scheme (Grant No. YSS/2015/000078) funding and the institute funding for the facilities setup at the Centre of Excellence, Materials Science/Sensors & Nanoelectronics at CMR Institute of Technology that was used for the conduction of this work.

References

[1]. S. Danfá, R.C. Martins, M.J Quina, J. Gomes, Supported TiO_2 in Ceramic Materials for the Photocatalytic Degradation of Contaminants of Emerging Concern in Liquid Effluents: A Review, Molecules 26 (2021) 5363. https://doi.org/10.3390/molecules26175363

[2]. M.T. David Maria, T. Antonella, S.S Andrijana, E. Leonardo, Effects of SiO_2 addition on TiO_2 crystal structure and photocatalytic activity, J. Eur. Ceram. Soc. 30 (12) (2010) 2481-2490. https://doi.org/10.1016/j.jeurceramsoc.2010.05.014

[3]. C. Kang, L. Jing, T. Guo, H. Cui, J. Zhou, H. Fu, Mesoporous SiO_2-Modified Nanocrystalline TiO_2 with High Anatase Thermal Stability and Large Surface Area as Efficient Photocatalyst. J. Phys. Chem. C, 113(3) (2009) 1006–1013. https://doi.org/10.1021/jp807552u

[4]. Z-Y. Yang, G-Y. Shen, Y-P. He, X-X. Liu, S-J. Yang, Preparation of TiO_2/SiO_2composite oxide and its photocatalytic degradation of rhodamine B. J. Porous Mater. 23 (2016) 589–599. https://doi.org/10.1007/s10934-015-0114-7

[5]. B. Czech, W. Buda, Photocatalytic treatment of pharmaceutical wastewater using new multiwall-carbon nanotubes/TiO_2/SiO_2 nanocomposites. Environ. Res. 137 (2015) 176–184. https://doi.org/10.1016/j.envres.2014.12.006

[6]. D. Lia, S. Risya, Application of Titanium-Silica-Graphite Composite Material for Photocatalytic Process of Methylene Blue, Indones. J. Chem. 20 (2020) 1271 – 1282. https://doi.org/10.22146/ijc.48998

[7]. X. Kang, S. Liu, Z. Dai, Y. He, X. Song, Z. Tan, Titanium Dioxide: From Engineering to Applications. Catalysts, 9 (2019) 191. https://doi.org/10.3390/catal9020191

[8]. S. Preetha, R. Pillai, S. Ramamoorthy, A. Mayeen, K. M. Archana, N. Kalarikkal, B. Narasimhamurthy, Surf. Interfaces 30 (2022) 101812. https://doi.org/10.1016/j.surfin.2022.101812

[9]. K. K. H. De Silva, H-H Huang, R. K. Joshi, M. Yoshimura, Chemical reduction of graphene oxide using green reductants, Carbon. 119 (2017) 190–199. https://doi.org/10.1016/j.carbon.2017.04.025

Prediction of Solar Radiation Using Data Driven Models

Nagaveni Thallapalli[1] and S V S N D L Prasanna[2*]

[1]Associate Professor, Department of Mechanical Engineering,
University College of Engineering(A), Osmania University, Hyderabad-500007, India.
[2]Associate Professor, Department of Civil Engineering,
University College of Engineering(A), Osmania University, Hyderabad-500007, India

prasanna.s@uceou.edu-*Corresponding author

Abstract. The local climatic conditions play a significant role in the amount of Solar Radiation that reaches the Earth's surface. In India, most of the locations, receive ample Solar Radiation. The exact evaluation of SR is most essential for engineering studies on energy and climate. For the present investigations, the meteorological data for Hyderabad city in Telangana state for the period 2009 - 2020 is adopted. Five climatological variables viz., Max. Temp. (T_{max}), Min. Temp. (T_{min}) and Mean Air Temp. (T_{mean}), Sunshine hours (n), and Relative Humidity (RH) were collected from the weather forecast. In the present investigations, two conventional empirical equations viz., Abdalla, and Angstrom are adopted for analytical estimation of Solar Radiation (SR). Furthermore, two data-driven models via., Artificial Neural Networks (ANN) and Multi Gene-genetic Programming (MGGP) are used to develop an equation for estimating Solar Radiation (R_s). The model equations were developed based on the parameters adopted for each conventional equation. These results are compared with conventional equation results. The present investigation exhibited that, the expected results of the Solar radiation of the arrangement formed by the data are favourable and also simplify the testing data. The corresponding R^2 values for random data are 0.976 for the training dataset, while the value for the testing dataset is 0.964. Based on the performance analysis for each equation, the efficiency and reliability of the proposed MGGP and ANN model are validated and predicted for the year 2021. The predicted results were found to be in good agreement with the analytical results evaluated from three properly used equations.

Keywords: Solar Radiation, Simulation, Multi Gene Genetic Programming (MGGP), Artificial Neural Network (ANN)

1. Introduction

The major key factor in the process of hydrological cycle and other energy cycles possible in the environment is Solar Radiation (SR)[1]. This parameter is the key in evaluation of Hydrological parameters which is critical for interpreting crop water requirements. Personnel involved in hydrological or water quality measurements must be generally familiar with the means of evaluating the SR values[2]. The authors investigated that, evapotranspiration values that are estimated for Southern France adopting Penman – Monteith method was observed to highly complex with respect to SR [3].

As per the RCP2.6 and RCP4.5representative concentration pathways, the SR values would regularly improve close to the level that conquered in the late twentieth century. They concluded that, it results from the peaking of aerosol loading before mid-century. Conversely, under RCP8.5, incident radiation would still decline, although more slowly than during the early century[4]. Most traditionally, the SR values are estimated using Satellite-based instruments since past. They stated that the higher SR values will be recorded under higher sunlight hours The investigators considered the values of SR and air temperature (Ta) as the key input data for estimating Evapotranspiration for adaptive neuro-fuzzy inference system (ANFIS) modelling[5]. Hatice Citakoglu and et al, proposed a new compact method using multi-gene genetic programming (MGGP) to estimate solar radiation in Turkey. 163 stations meteorological data from 1975 to 2015 for7 areas of Turkey were used for MGGP modelling. It was concluded that the MGGP modelling and empirical equations obtained after calibration, are found to most effective for estimating Solar Radiation[6].

Multiple Genetic Programming (MGP) technique projected by Ali Danandeh Mehr et al., stated that, this technique shows an improvement in the accuracy of prediction when compared to the standalone Genetic Programming (GP). The measurement of the amount of Rainfall from two weather forecasting stations in the Lake Urmia Basin, Iran, are used to demonstrate the model's 1-month rainfall forecasting capability. They concluded that the MGP model outstrips the standards at both the station points in terms of statistical performance[7]. Anurag Malik et al. investigated the ability of various predictive models, comprising of Support Vector Machine (SVM), Multiple Model-Artificial Neural Network (MMANN), Multivariate Adaptive Regression Spline (MARS), Multi-Gene Genetic Programming (MGGP), and 'M5Tree,' to estimate pan evaporation on a monthly scale (EPm) at two stations in India (e.g., Ranichauri and Pantnagar). The Gamma test was used to determine which input variables were the most effective for the five different models. According to the findings, MM-ANN-1 and MGGP-1 models will aid local participants in the proper management of water resource[8].

Many methods are implemented for the prediction of Solar Radiation via empirical equations. In the existing state, the application of data-driven models is gaining importance in predicting various hydrological parameters. Garg et al have proposed a simulation approach (MGGP) for the relation amongst water content and two input parameters via., soil suction and volumetric crop root content. The projected method employs a statistical stepwise regression approach[9].

In the present study, a total of 156 data points that include6input variables via., Max. Temp., Min. Temp. (°C), Mean Air Temp (T_{mean}) (°C), Relative Humidity (%), Sunshine hours were collected from the meteorological sites[10]. The Radiation (Ra) Values were computed from the tables. The objectives of the present investigations are,

- Analytical estimation of Solar Radiation (SR) adopting two conventional empirical equations via., Angstrom and Abdalla.
- To develop model equation, estimating Solar Radiation (R_s) for two methods using two data-driven models via., Artificial Neural Networks (ANN) and Multi Gene-genetic Programming (MGGP). The model equations were developed based on the parameters adopted for each conventional equation.
- Further, the study was extended to compare the statistical parameters obtained from ANN and MGGP model.

Although there have been numerous applications of data-driven models in hydrological studies, this study is an attempt to incorporate ANN and MGGP to test the estimation accuracy of Solar Radiation values.

2. Data Collection

The study area comprises of Hyderabad city in Medchal district from Telangana State. It is the capital of Telangana State is Hyderabad, located on the banks of the Musi River and the Deccan Plateau. Hyderabad is located in central Telangana and is spread over an area of 2,500 km^2. The city lies in the Deccan Plateau and rises to an average height of 536 m above sea level. The city lies at 17.366° N latitude and 78.476° E longitude. Major parts of the city feature rocky terrain and, in some parts hills, can also be noticed. Figure 1 depicts the map of the study area location (https://en.wikipedia.org)

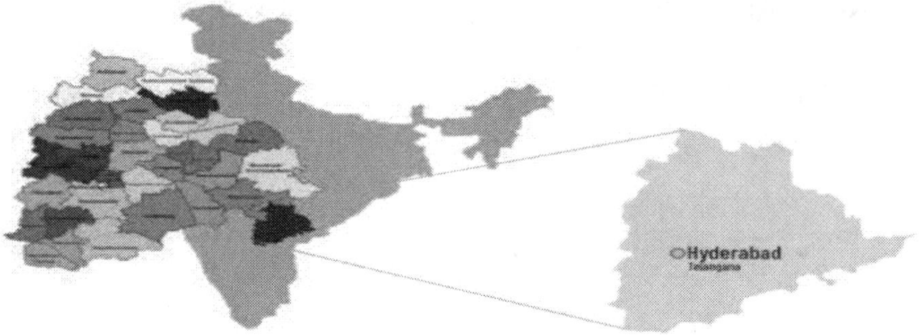

Figure 1. Study Area.

3. Methodology

The present study is carried out using a two-way methodology via., analytical calculation and mathematical modelling analysis. A detailed understanding is present in the subsequent sub-sections.

3.1. Analytical Analysis

The Solar Radiation (SR) is evaluated as a part of Penman-Monteith method for the estimation of Evapotranspiration (ETo). This parameter plays a significant role in the ETo estimation[10]. In order to estimate SR values, the following two equations are adopted via., Angstorm, and Adballa in the present study. The input parameters via., Max. Temp., Min. Temp. (°C), Mean Air Temp (T_{mean}) (°C), Relative Humidity (%), Sunshine hours (n) were considered on monthly basis for 13 years. The redolent statistical parameters of the meteorological variables are presented in Figure 2. The Solar Radiation (R_s) values calculated using Angstorm highlighted in equation (1)[11]. Abdalla(1994) recommended a new equation for SR estimation from sunshine durations (n), maximum monthly mean sunshine durations (N), mean temperature, and relative humidity, shown in equation (2)[12]

$$R_s = \left(a + b * \frac{n}{N}\right) * R_a \tag{1}$$

$$R_s = \left(a + b * \frac{n}{N} + c * T_{mean} + d * RH_{mean}\right) * R_a \tag{2}$$

where, Rs is solar radiation (MJ/m^2), Ra is extra-terrestrial radiation (MJ/m^2), n is sunshine durations; N is maximum monthly mean sunshine durations; and a and b are coefficients whose values as suggested in area = 0.25, b = 0.50, T_{max} is maximum long-term monthly mean temperature(°C), T_{min} is minimum long-term monthly mean temperature(°C), and a is a coefficient, for which the numerical values of 0.19 for coastal zones and 0.17 for terrestrial zones were recommended, T_{mean} is monthly mean temperature; RH is monthly mean moisture; and a, b, c, d are coefficients (a = 1.943, b = 0.577, c = −0.01483, and d = −0.12129).

3.2. Multi Gene Genetic Programming (MGGP)

Multi Gene Genetic Programming (MGGP) is a newly developed method of Genetic Programming (GP)[13]. It mainly helps in the improvement of the precision of GP. MGGP mainly differs from GP in the number of trees that can be used in MGGP. GP mainly uses single trees whereas MGGP uses a combination of trees. A typical MGGP model is shown in Figure 3. The advantage of the MGGP process is that it is a data-driven procedurethat builds models using experimental data. In MGGP the final structure of the target model is made in a linear form by combining two or more non-linear genes. The general form of a MGGP model is shown in equation (3).

Name of the Equation	Input Parameters	Data Points	Mean	Standard Deviation	Skewness
Angstrom and Abdalla Equations	Mean Air Temp (°C)	Training	27.667	3.637	0.962
		Testing	27.513	3.729	1.217
	Relative Humidity (%)	Training	52.179	17.280	0.031
		Testing	55.538	18.122	0.255
	Average Sunlight hrs, n	Training	11.341	0.922	-1.620
		Testing	10.393	1.862	-1.395
	Mean daylight hours, N	Training	12.156	0.709	-0.352
		Testing	12.080	0.716	0.225
	R_a (MJ/m²/day)	Training	35.013	4.263	-1.156
		Testing	34.554	4.346	-0.891
	Solar Radiation (Angstrom)	Training	24.997	2.155	-0.260
		Testing	23.388	2.893	-0.007
	Solar Radiation (Abdalla)	Training	70.029	5.837	-1.430
		Testing	67.592	5.981	-0.178

Figure 2. Statistical Parameters of Input

$$y = a_0 + a_1 Gene_1 + a_2 Gene_2 + \cdots + a_n Gene_n \qquad (3)$$

where, y = output, a_0 = bias term, $Gene_i$ = non-linear genes, a_i= coefficient of related genes, n= number of genes.

This avoids errors due to differences that are often used when building a model of the differential equation. Then a method of least squares is used in the MGGP algorithm to find the coefficients $a_0, a_1, \ldots\ldots\ a_n$. This procedure reduces the error percentage caused by different variables that are frequently used when developing a differential equation model. The models are presented as a tree structure, along with the generation of the huge population of these models. Each case has a terminal and a function set. The basic mathematical operators (+, -, /), Boolean algebra operators (e.g., AND and OR), or other user-defined mathematical symbols form the elements of the function set. The problem with the input variables represents the terminal set [14].

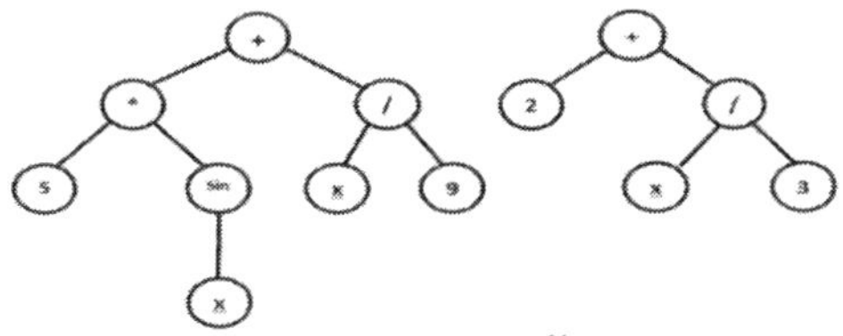

$$y = a_0 + a_1 \times \left[5 \times (Sin(x)) + \left(\frac{x}{9}\right)\right] + a_2 \times \left[2 + \left(\frac{x}{3}\right)\right]$$

Figure 3. Typical MGGP Model.

In this work, the software tool GPTIPS is used to develop the equations for the prediction of Solar radiation[15]. This software is a new code written by "Genetics and Regression Symbolic Programming" based on multi-gene GP used with MATLAB 2021a.

The standards of the maximum number and the depth of the genes determine the extent and different systems of the model for the space of universal solutions. For the present study, out of the 156 data points obtained from 13 years of data, 117 points were selected for training and 39 points for testing. The input parameters considered for the development of the MGGP and ANN model based on three different empirical equations are highlighted in Table -1.

The individual performance fitness functions are evaluated via. R^2, Mean Square Error (RMSE), MAPE and MAE are given as follows:

$$R^2 = \left[\frac{\sum_{i=1}^{n}(a_i - \bar{a}_i)(p_i - \bar{p}_i)}{\sqrt{\sum_{i=1}^{n}(a_i - \bar{a}_i)^2 \sum_{i=1}^{n}(p_i - \bar{p}_i)^2}}\right]^2 \qquad (4)$$

$$\mathrm{RMSE} = \sqrt{\frac{1}{\mathrm{n}}\sum_{i=1}^{\mathrm{n}}(a_i - p_i)^2} \tag{5}$$

$$MAPE = \frac{\sum_{i=1}^{n}|a_i - p_i|}{\sum_{i=1}^{n} a_i} \times 100 \tag{6}$$

$$MAE = \frac{\sum_{i=1}^{n}|a_i - p_i|^2}{n} \tag{7}$$

where, n = total number of data points, a_i&p_i= actual output values, $\bar{a}_i$ and $\bar{p}_i$= average predicted and actual outputs.

Table 1. Detail of Input Parameters.

Name of the Equation	Input Parameters for Training and Testing					
	x_1	x_2	x_3	x_4	x_5	y
Angstorm	Avg. Sunlight hours (n)	Mean daylight hours (N)	Extra-terrestrial radiation (Ra)	--	--	Solar Radiation (R_s)
Abdalla	Mean Air Temp	Mean Relative humidity (%)	Avg. Sunlight hours (n)	Mean daylight hours (N)	Extra-terrestrial radiation (Ra)	

The higher value of R^2 and lower values of MAE, MAPE and RMSE are the signs of developed equations having better performance. The other initial parameters were established for the implementation of GPTIPS is given in Table 2.

Table 2. Parameter Settings for MGGP.

Run Parameter	Value for MGGP	Value for ANN
Population size/ Number of epochs	250	1000
Max. generations	150	--
Input variables	3 - Angstorm and 5-Abdalla	--
Training instances	117	117
Testing Instances	39	39
Tournament size	25	--
Elite fraction	0.7	--
Probability of pareto tournament	0.7	--
Max. genes/ Number of neurons in hidden layers	6	1-10
Max. tree depth/ No of hidden layers	4	2
Max. total nodes	Inf	--
Crossover probability	0.84	--
Mutation probabilities	0.14	--
Complexity measure	Expressional	--
Function set	×, -, +, tanh, multiply 3 variables and add three variables	Tagsig - Logsig-Pureline

4. Results and Discussions

The present section expresses the results assessed after processing the 13 years of data.

- The average SR values obtained from two empirical equations via., Angstorm and Abdalla equations for 156 data points (2009 – 2021) are depicted in Figure 4.
- The MGGP and ANN model equations were developed for two different traditional equations via., Angstorm and Abdalla equations for 156 data points.
- The following equation (8), and equation (9) were evolved as the best model for the estimation of Solar radiation as per Angstrom, and Abdalla equations respectively.

Angstrom:

$$y = 0.00325\ x_1 + 0.109\ x_2 + 0.215\ x_3 - 0.00325\ \tanh(x_2) + 0.00992\ x_2^2\ \tanh(x_2) + 0.00277\ x_1\ (x_1 + 2.0\ x_3) - 0.00325\ x_2^2 - 0.0755\ \tanh(x_1^2)\ \tanh(x_2 + x_3)\ (x_1 + 2.0\ x_2 - 1.0\ x_1\ x_3) - 0.00325\ x_1\ x_2\ x_3 + 1.07 \quad (8)$$

Abdalla:

$$y = 1.68\ x_5 - 0.15\ x_1 + 0.0519\ x_3\ x_5 - 9.35e^{-6}\ x_2^2\ x_3 - 1.9e^{-5}\ x_1\ x_4\ x_5^2 - 9.6e^{-5} x_1\ x_2\ x_3 + 4.77 \quad (9)$$

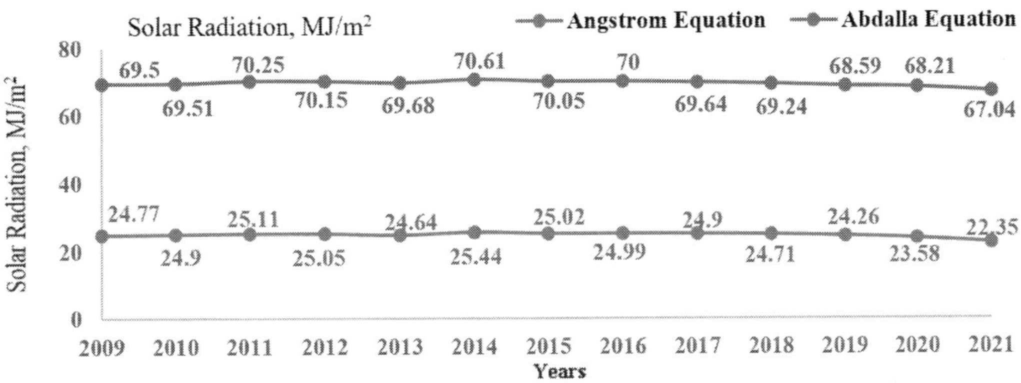

Figure 4. Average Analytical values of SR.

- The accuracy of the equation can be elaborated by plotting the measurement obtained from training and test data. Figure 5 depicts the measured SR values (expected output) and the one calculated by MGGP (predicted output), demonstrating an excellent correlation between the multi gene GP predictions and the measured data.
- The comparison between the actual and predicted values of SR is depicted in Figure 5. The bias terms in the selected cross-validated mathematical model were estimated as +1.07 for Angstorm and +4.77 for Abdalla equations respectively, as detailed in equation (7) and (8).
- The determination of the coefficient values of R^2 - goodness of fit, RMSE - root mean squared error, MAE - mean absolute error, SSE - sum of squared errors, and MSE - mean squared errors are highlighted for the three empirical equations in Table 3.

Table 3. Metrics for Evaluation for Empirical equations.

Metric	Angstorm Equation		Abdalla Equation	
	Training	Testing	Training	Testing
R^2	0.99973	0.99974	0.99984	0.99915
RMSE	0.035661	0.046394	0.074387	0.17479
MAE	0.028392	0.040092	0.056338	0.11925
SSE	0.14879	0.083943	0.64742	1.1915
Max. abs. error	0.080445	0.09449	0.23104	0.74244
MSE	0.0012717	0.0021524	0.0055335	0.030551

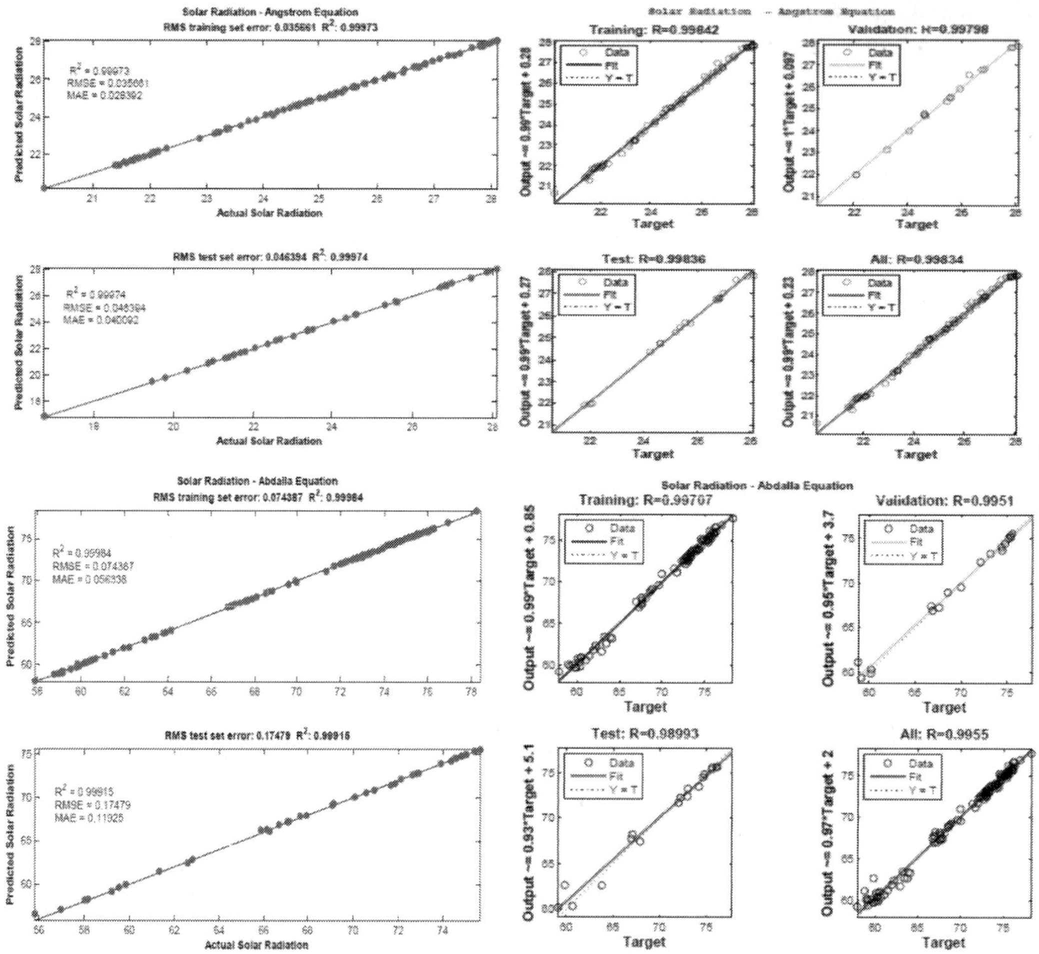

Figure 5. Graphs of Measured and Predicted Values of Evapotranspiration by MGGP & ANN model.

5. Conclusions

In the present paper, Solar Radiation values were estimated for Hyderabad city of Telangana State. The meteorological data adopted for the study was collected on a monthly basis for 13 years. Total 117 data points were adopted for the training and 39 data points for testing out of 156 data points. For analytical analysis, two widely adopted equations via., Angstorm and Abdalla were used to evaluate the Solar Radiation. Numerical modelling is carried out using MGGP and ANN methods and compared based on the statistical parameters via., mean standard deviation, and skewness. Further, the commonly adopted performance criteria of MAE, MSE, SSE, RMSE, and R^2 statistics were also compared. The following conclusions have been are drawn after the critical observation of the results:

- The prediction result of the neural network model which has 22 neurons in hidden layer was found to be in good agreement with the experimental data.
- The distribution of data points for neural network model and MGGP models are almost similar and close to the actual experimental data with correlation coefficient (R) in range in range of 0.99 - 1.00.
- The estimation performance of the MGGP model developed equations, for two equations via., Angstorm and Abdalla were compared and found that both the datasets exhibited good consonance.

- The higher values of multiple regression coefficients evaluated from both the equations indicate that, the developed model can predict the physical phenomena more accurately in good consonance with the conventional methods.
- From the analysis, based on the R^2 value it can concluded that, mathematical modelling can be conveniently adopted for the estimation of Solar Radiation.

Thus, the present work emphasizes the use of numerical modelling for the accurate and feasible estimation of Solar Radiation to choose optimal values depending on their needs. The proposed models MGGP and ANN can effectively integrate the effects of various parameters. The ability of MGGP and ANN to generate powerful compact models can be concluded as a significant advantage. The model can be extended with the addition of a greater number of governing factors and also past data sets, to further check the consistency and applicability in the prediction of Solar Radiation.

6. References

[1] Kalogirou S A, Costas Neocleous, Stelios Pashiardis, and Christos Schizas 1999 Wind speed prediction using artificial neural networks, *Atmospheric Sciemces.*

[2] Bartram J and R Ballance 1996 Water quality monitoring: a practical guide to the design and implementation of freshwater quality studies and monitoring programmes, *CRC Press.*

[3] Bois B et al 2008 Using remotely sensed solar radiation data for reference evapotranspiration estimation at a daily time step, *Agricultural and Forest Meteorology*,148 p. 619-630.

[4] Ruosteenoja K 2019 Future changes in incident surface solar radiation and contributing factors in India in CMIP5 climate model simulations, *Journal of Applied Meteorology and Climatology*,. 58(1): p. 19-35.

[5] Cobaner M 2011 Evapotranspiration estimation by two different neuro-fuzzy inference systems, *Journal of Hydrology*, 398(3-4): p. 292-302.

[6] Citakoglu H, B Babayigit, and N A Haktanir, 2020 Solar radiation prediction using multi-gene genetic programming approach, *Theoretical and Applied Climatology*, 142(3): p. 885-897.

[7] Mehr A D and M J S Safari 2020 Multiple genetic programming: a new approach to improvegenetic-based month ahead rainfall forecasts, *Environmental Monitoring and Assessment*, 192(1): p. 1-12.

[8] Malik A 2020 Modeling monthly pan evaporation process over the Indian central Himalayas: application of multiple learning artificial intelligence model, *Engineering Applications of Computational Fluid Mechanics*, 14(1): p. 323-338.

[9] Garg A L, Rachmawati and K Tai 2013 Classification-driven model selection approach of genetic programming in modelling of turning process, *The International Journal of Advanced Manufacturing Technology*, 69(5-8): p. 1137-1151.

[10] Allen R G et al 1998 Crop evapotranspiration-Guidelines for computing crop water requirements-FAO Irrigation and drainage paper 56. *FAO*, Rome, 300(9): p. D05109.

[11] Angstrom A 1924 Solar and terrestrial radiation. Report to the international commission for solar research on actinometric investigations of solar and atmospheric radiation, *Quarterly Journal of the Royal Meteorological Society,* 50(210): p. 121-126.

[12] Abdalla Y A G 1994 *New Correlations of Global Solar Radiation with Meteorological Parameters For Bahrain,* *International Journal of Solar Energy*, 16(2): p. 111-120.

[13] Koza J R et al 1999 *Genetic programming III: Darwinian invention and problem solving*. 3, Morgan Kaufmann.

[14] Garg A, K Tai, and M 2014 Savalani Formulation of bead width model of an SLM prototype using modified multi-gene genetic programming approach, *The International Journal of Advanced Manufacturing Technology*, 73(1-4): p. 375-388.

[15] Searson, D P, D E Leahy, and M J Willis 2010 GPTIPS: an open source genetic programming toolbox for Multigene Symbolic Regression, i*n Proceedings of the International multiconference of engineers and computer scientists.* Citeseer.

Development of Machine Learning models for air pollutants using weka

S Venkatramana Reddy[1], N Madhavi Latha[2] and B. Sarojamma[3]*

1: Department of Physics, Sri Venkateswara University, Tirupati, A.P., India.

2&3: Department of Statistics, Sri Venkateswara University, Tirupati, A.P., India.

* Author for Correspondence e-mail: saroja14397@gmail.com

Abstract. All over the world desecration is a major factor for livelihood and plantation. There are various types of desecrations and major are air desecration (or desecration), water desecration or water desecration, Industrial desecration, noise desecration, urban desecration etc. In India Act 1981 prescribes various functions for the Central Desecration Control Board (CPCB) at the open level and state desecration control board at the state level about air desecration. Act 1974 gives about water prevention and control desecration. Act 1986 section 2(a) explains noise desecration under rules, 2000. In this paper we are collected data of air pollutants such as particulate matter of 2.5 microns or less. particulate matter of 10 microns or less, Nitrogen Monoxide, Nitrogen dioxide, chemical compound of oxygen and nitrogen, Azane or Ammonia, Carbon Monoxide, Sulfur dioxide Ozone,Benzene, Toluene, Xylene, AQI, AQI_Bucket. For these variables we are fitted Machine Learning models such as Naive Bayes Classifier, CV Parameter Selection, K-nearest neighbours, J48 and Hyper Pipes using WEKA software. Detailed accuracy for models like TP ratio, FP ratio, precision, F-measure, ROC area are estimated. Using Root average square error, Average absolute error, we examine which is the best model.
Keywords: air pollutants, Naive Bayes Classifier, CV Parameter Selection, Multi class Classifier, J48, and Hyper Pipes, RMSE.

1. Introduction

Desecration means effecting the land, water, air or other parts of the environment dirty and not safe or suitable to use. There are various types of desecrations in the literature, they are air desecration, light desecration, litter, noise desecration, plastic desecration, Soil contamination, radioactive contamination, thermal desecration, visual desecration and water desecration. Air desecration is due to release of chemicals and particles into atmosphere, light trespass, over-illumination and astronomical interference is light desecration, Road noise, aircraft noise, industrial noise cause noise desecration, discharge of industrial waste water, discharge of untreated Sewage and chemical contaminants etc., called as water desecration.

Begum Ligsar and DenizUnal [1] in their research paper discussed about the banking analysis and insurance using data mining techniques. They divide total study into two parts, primary part explains about to selection of the best model for data and secondary part describes the best model for analysis using ODD's ratio. Larissa E. Back et al [2] explains in research paper that it contains seven sections, first section explains about research articles related to wind speed and their importance, second part describes data description used by authors, third section explains about relationship between daily mean wind speed with precipitation, fourth part describes about uncertainties involved in precipitation, fifth section explains about correlation between precipitation and wind speed, sixth section discusses changes in + and – signs about wind speed growth and decline, and last section explains summary and conclusions. Chinnai yan ponnuraja et al. [3] explains machine learning techniques for infected tuberculosis patients' data. Hari Mallikarguna Reddy et al.[4] describes fitting of wind speed data using different data mining techniques and for choosing best data mining algorithm using different measures of accuracy. Geetha et al. [5] took the atmospheric data of day wise for the 5 years from 2014 and applying data mining algorithms using WEKA. Ghogare and Kalmegh [6] discussed about performance comparison of Random forest and Hoeffding Tree classifier using WEKA on Car reviews data. Ajay Kumar Mishra and Bikram Kesari Ratha [7] published a paper on study of Random Tree and Random Forest data Mining Algorithms for Micro array data Analysis.

2. Methodology

We took the Day wise data of Desecration Components like particulato matter of 2.5 microns or less, particulate matter of 10 microns or less, Nitrogen Monoxide, Nitrogen dioxide, chemical compound of oxygen and nitrogen, Azane or Ammonia, Carbon Moxoxide, Sulfur dioxide, Ozone, Benzene, Toluene, Xylene and AQL of Visakhapatnam of Andhra Pradesh, India from 7.1.2016 to 7.1.2020

Expansions:

PM2.5- particulate matter of 2.5 microns or less

PM 10- particulate matter of 10 microns or less

NO- Nitrogen Monoxide

NO_2- Nitrogen dioxide

NO_X- chemical compound of oxygen and nitrogen

NH_3 - Azane or Ammonia

CO- Carbon Moxoxide

SO_2- Sulfur dioxide

O_3 - ozone

Benzene

Toluene and Xylene

By using WEKA Software, we choose Naive Bayes classifier, CV parameter selection, J48, K-nearest neighbours and Hyper pipes Techniques.

Naive Bayes classifier: Naive Bayes algorithm is dependening upon probabilistic models. In data mining techniques using WEKA having below flow chart.

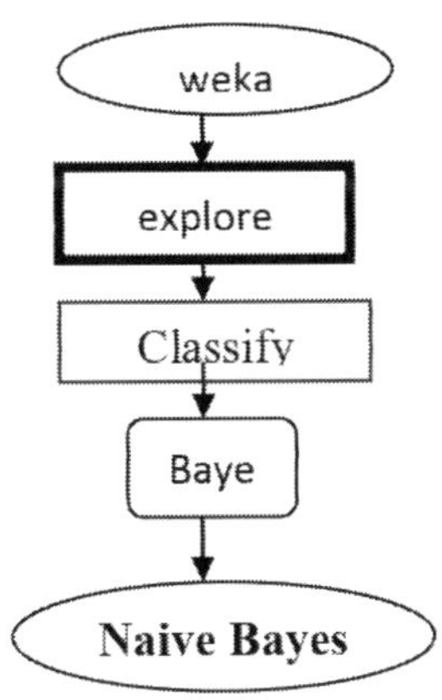

Journal of Physics: Conference Series **2332** (2022) 012018 doi:10.1088/1742-6596/2332/1/012018

Hyper pipe line: Hyper pipe line is estimated using the following flow chart

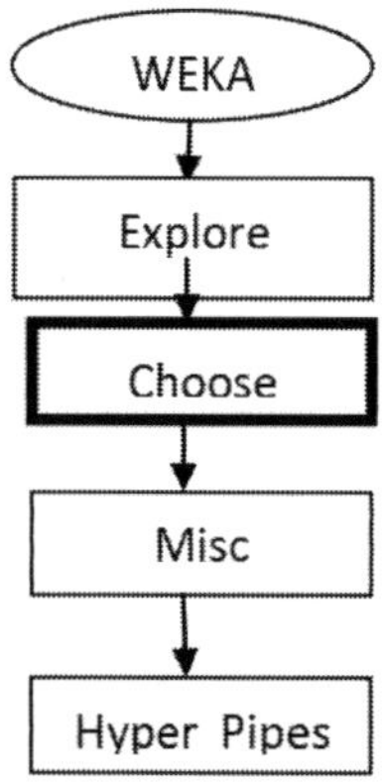

CV parameter selection: Flow chart for estimation of CV parameter is as follows

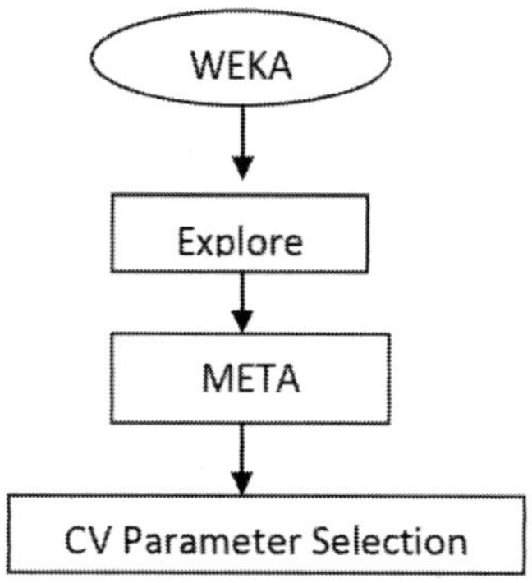

J48: It is also an Decision tree with allowing missing values. Decision tree of J48 is also called C4.5. Flow chart for estimation of J48 is

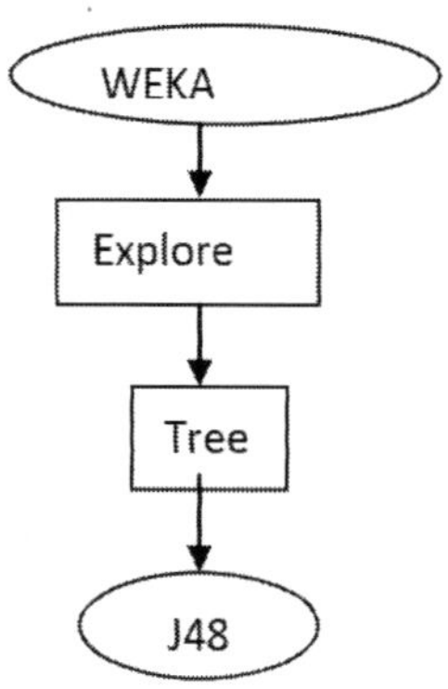

K-nearest neighbors: Flow chart for K nearest neighbor classifier **is as follows**

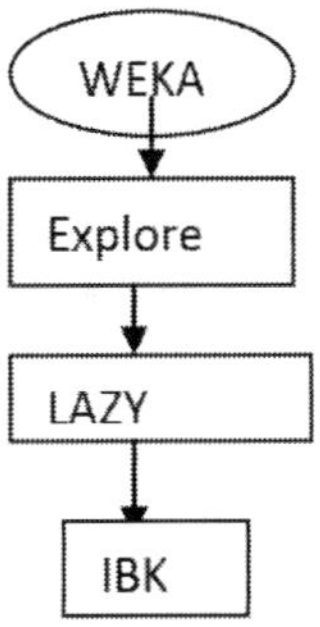

3. Empirical Investigations:

In this paper we are fitting Naive Bayes Algorithm, Multi class classifier, K-nearest neighbours, J48 and hyper pipes models using WEKA software. For day wise AQI data of Vishakhapatnam, in Andhra Pradesh, India, Air quality index measurements like particulate matter at 2.5 and 10, Nitrogen Monoxide, Nitrogen dioxide, ammonia, carbon monoxide, sulfur dioxide, Ozone , Benzene, Toluene and Xylene from 7.1.2016 to 7.1.2020 for analysis [8].

By using WEKA software, Naive Bayes classifier gave True positive ratio (TPR), False Positive ratio (FPR), Precision (PR), Recall, F-Measure, Receiver operating characteristic Area (ROCA) for classes like attribute Satisfactory, Moderate, Good, Poor, Very Poor are listed in table 1.

Table 1. Naive Bayes classifier Medthod

class	TPR	FPR	PR	Recall	F Measure	ROCA
Satisfactory	0.83	0.128	0.8	0.83	0.814	0.915
Moderate	0.8	0.078	0.91	0.8	0.851	0.909
Good	0.981	0.028	0.622	0.981	0.761	0.989
Poor	0.932	0.027	0.697	0.932	0.798	0.989
Very poor	0.667	0.003	0.75	0.667	0.706	0.994

By using classifiers rules zero R for CV parameter selection of data mining techniques using WEKA software for Air quality index measurements like particulate matter at 2.5 and 10, Nitrogen Monoxide, Nitrogen dioxide, ammonia, carbon monoxide, sulfur dioxide, Ozone , Benzene, Toluene and Xylene from 07.01.2016 to 07.01.2020 of Visakhapatnam gave TP ratio, FP ratio, precision, Recall, F Measure and ROC area are listed in table 2

Table 2

Table 2. CV parameter selection model

class	TPR	FPR	PR	Recall	F Measure	ROCA
Satisfactory	0	0	0	0	0	0.496
Moderate	1	1	0.405	1	0.662	0.5
Good	0	0	0	0	0	0.485
Poor	0	0	0	0	0	0.483
Very poor	0	0	0	0	0	0.457

Using Hyper pipe line by 10 fold cross validation for Air quality index measurements like particulate matter at 2.5 and 10, Nitrogen Monoxide, Nitrogen dioxide, ammonia, carbon monoxide, sulfur dioxide, Ozone , Benzene, Toluene and Xylene from 07.01.2016 to 07.01.2020 of Visakhapatnam explains TP ratio, FP ratio, precision, Recall, F Measure and ROC area are listed in table 3.

Table 3. Hyper pipe line method

Class	TPR	FPR	PR	Recall	F Measure	ROCA
Satisfactory	0.993	0.029	0.955	0.993	0.974	0.823
Moderate	0.995	0.034	0.966	0.995	0.98	0.714
Good	0.692	0	1	0.692	0.818	0.993
Poor	0.851	0.001	0.984	0.851	0.913	0.878
Very poor	0.5	0	1	0.5	0.667	0.988

K-nearest neighbours Using 10 fold cross validation for Air quality index measurements like particulate matter at 2.5 and 10, Nitrogen Monoxide, Nitrogen dioxide, ammonia, carbon monoxide, sulfur dioxide, Ozone , Benzene, Toluene and Xylene from 07.01.2016 to 07.01.2020 of Visakhapatnam explains TP ratio, FP ratio, precision, Recall, F Measure and ROC area are listed in table 4.

Table 4. K-nearest neighbours model

class	TPR	FPR	PR	Recall	F Measure	ROCA
Satisfactory	0.801	0.141	0.778	0.801	0.789	0.845
Moderate	0.838	0.149	0.847	0.838	0.842	0.72
Good	0.577	0.014	0.652	0.577	0.612	0.79
Poor	0.703	0.018	0.722	0.703	0.712	0.842
Very poor	0.611	0.007	0.579	0.611	0.595	0.784

J48 Using 10 fold cross validation for Air quality index measurements like particulate matter at 2.5 and 10, Nitrogen Monoxide, Nitrogen dioxide, ammonia, carbon monoxide, sulfur dioxide, Ozone , Benzene, Toluene and Xylene from 07.01.2016 to 07.01.2020 of Visakhapatnam explains TP ratio, FP ratio, precision, Recall, F Measure and ROC area are listed in table 5.

Table 5. J48 model

class	TP Ratio	FP Ratio	Precision	Recall	F Measure	Roc Area
Satisfactory	1	0	1	1	1	1
Moderate	0.998	0	1	0.998	0.999	0.999
Good	1	0	1	1	1	1
Poor	0.986	0.001	0.986	0.986	0.986	0.991
Very poor	1	0.001	0.947	1	0.973	1

Different Measures of accuracy like, Kappa statistic, Mean absolute error, Root Mean Squared error, Relative Absolute error and Root relative squared error, for Naive Bayes classifier, CV parameter, Hyper pipe line, K-nearest neighbours and J48 by 10 fold cross validation Air quality index measurements like particulate matter at 2.5 and 10, Nitrogen Monoxide, Nitrogen dioxide, ammonia, carbon monoxide, sulfur dioxide, Ozone, Benzene, Toluene and Xylene for day wise data of 07.01.2016 to 07.01.2020 of Vishakhapatnam, Andhra Pradesh, India are listed in table-6

Table 6. Accuracy measures

Method	Kappa statistic	MAE	RMSE	RAE	RRSE
Naive Bayes classifier	0.00	0.2415	0.3472	100	100
CV parameter	0.7502	0.2984	0.3738	123.5924	107.6665
Hyper pipe line	0.9393	0.2929	0.3704	121.2738	106.6941
K-nearest neighbours	0.6677	0.0811	0.2821	33.5704	81.236
J48	0.9972	0.0007	0.0261	0.2829	7.5279

4. **Summary and Conclusions**:

Estimation of Air quality Index for Vishakhapatnam, Andhra Pradesh, India as indicator of Satisfactory, Moderate, Good, Poor and Very poor by using data mining techniques such as Naive Bayes classifier, CV parameter, K-nearest neighbours, J48 and hyper pipeline with the help of WEKA Software. AQL parameters like PM2.5, PM10, Nitrogen Monoxide, Nitrogen dioxide, Ammonia, Carbon Monoxide, Sulfur dioxide, Ozone, Benzene, Toluene, Xylene and AQI Indicator are estimated using Data mining Techniques. Different measures of accuracies like Kappa statistic, Mean absolute error, Root Mean squared error, Relative Absolute Error and Root relative squared error for the Models are listed in Table 7.

TABLE 7. Measures of Errors

Method	Naive Bayes classifier	CV parameter	Hyper pipe line	K-nearest neighbours	J48
MAE	0.2415	0.2984	0.2929	0.0811	0.0007
RMSE	0.3472	0.3738	0.3704	0.2821	0.0261
RAE	100	123.592	121.274	33.5704	0.2829
RRSE	100	107.667	106.694	81.236	7.5279

By observing Root Mean squared error values, best model among Naive Bayes classifier, CV parameter, K-nearest neighbours, J48 and hyper pipeline models is J48.

References:

[1] Cigsar B and Unal D 2019 Comparison of Data Mining Classification Algorithms Determining the Default Risk, Hindawi Scientific Programming **2019** 8706505.

[2] Larissa E. Back and Christopher S. Bretherton 2005 The Relationship between Wind Speed and Precipitation in the Pacific ITCZ, J. Climate **18** 4317.

[3] Ponnuraja C, Babu C Lakshmanan, Srinivasan V 2016 Performance Accuracy between Classifiers in Sustain of Disease Conversion for Clinical Trial Tuberculosis Data: Data Mining Approach, IOSR Journal of Dental and Medical Sciences **15** 105.

[4] Hari Mallikarjuna Reddy B, Venkatramana Reddy S, and Sarojamma B 2021 Data Mining Techniques for estimation of wind speed using WEKA, Int J. of Computer Sciences and Engineering **9** 49.

[5] Geetha K, Venkatramana Reddy S and Sarojamma B 2021 Comparative study of Decision Trees for Weather Data, Int J. of Engineering and Science Invention **10** 01.

[6]. Ghogare S A and Kalmegh S R 2019 Performance Comparison of Random Forest and Hoeffding Tree classifier using WEKA data mining tool on Car reviews data, International Journal of Engineering Research and Application **9** 37.

[7] Ajay Kumar Mishra, Bikram Kesari Ratha 2016 Study of Random Tree and Random Forest Data Mining Algorithms for Microarray Data Analysis, International Journal on Advanced Electrical and Computer Engineering **3** 5.

[8] Website; https://www.kaggle.com/#

A review of Economic Order Quantity modelling, their extensions and applicability

Uma Raju

Assistant Professor, Department of Mathematics,

CMR Institute of Technology, Bengaluru, India

Abstract

Inventory control and management is an important task for every organization, wherein they need to determine the order quantity and order period to maintain enough inventory to meet the demand. The original Economic Order Quantity (EOQ) model was developed by Ford W. Harris to address this problem and determine the optimal order quantity that helps in optimizing the cost incurred by the organization. Even though this model gives a very good approximation on these parameters, there were many inherent restrictions to the model that were based on un-realistic expectations. In over a century since its original publication, many researchers have made extensions to this model to address the limitations and align it with real world scenarios. In this review paper, we intend to provide a representation of the original EOQ model, its limitations and the various extensions made to the model to address these limitations.

Keywords: EOQ, Inventory cost, Holding cost, Perishable items, Delayed payments, Lead time, Replenishment, Price discount, Backorder, Backlog

Introduction

The material cost contributes a significant portion to the total cost of a manufactured product, and any additional inventory held by the company locks up its capital from being put to better use. Given these, it is very essential that the Inventory Management and control of an organization should be a top priority for the management. This planning involves things such as how much quantity of raw material should be ordered and at what time or frequency for optimal cost structures. Economic order quantity (EOQ) model, which was originally published by Ford W. Harris (1913) [1] and extensively studied and applied by R.H. Wilson (1934) [2] gives an optimal quantity to order which minimizes the total cost to the organization. This classical EOQ model has served as a basis for multiple extensions with time to address varying conditions and specific requirements.

In this paper, we will provide a basic representation of the classical EOQ model, the limitations of the classical model and subsequently discuss a few extensions to this model that were developed for specific requirements and practical scenarios.

1. Classical EOQ model

There are three types of cost associated with inventory management in general-

i) Ordering Cost (Co) – New orders need to be placed with the supplier based on the requirements, and there are multiple activities like supplier selection, price negotiation, purchase order generation, receipt of material etc. Each of these activities have an associated cost which makes up the ordering cost. For large quantity orders, we need fewer orders and hence lower Co and vice versa.

ii) Holding Cost (Ch) – This includes the cost of storing the inventory, insurance cost, obsolescence cost, pilferage and wastage etc. These costs make up the Holding cost. The Holding costs usually increase linearly with the inventory level.

iii) Shortage Cost – This cost results from shortage in inventory, which can result in lost sales, backorder, customer dissatisfaction etc.

The Classic EOQ model was made with the following assumptions –

i) A deterministic and constant annual demand of D
ii) A constant and known Lead time – L
iii) The ordering cost is constant and does not depend on order size – Co
iv) No limit on order size either for the supplier or the firm
v) Constant cost of holding a unit – Ch (per unit).
vi) Fixed unit cost of the item, does not change with time or order quantity – p
vii) No stockouts are allowed and all the demand is met.

With the above assumptions, the total cost of the plan (TC) is given by,

$$TC(Q) = \text{Total Holding Cost} + \text{Total Ordering cost, i.e.,}$$

$$TC(Q) = ((Q/2)*Ch) + ((D/Q)*Co))$$

Here Q is the quantity per order and Q/2 is the average holding per cycle. The optimal or EOQ (Q*) is determined by finding the minimum of the Total cost, which is given by

$$Q^* = \sqrt{((2Co*D)/Ch)}$$

and the optimal number of orders is given as (D/Q*).

A graphical representation of the cost and the EOQ is as given in Fig. 1

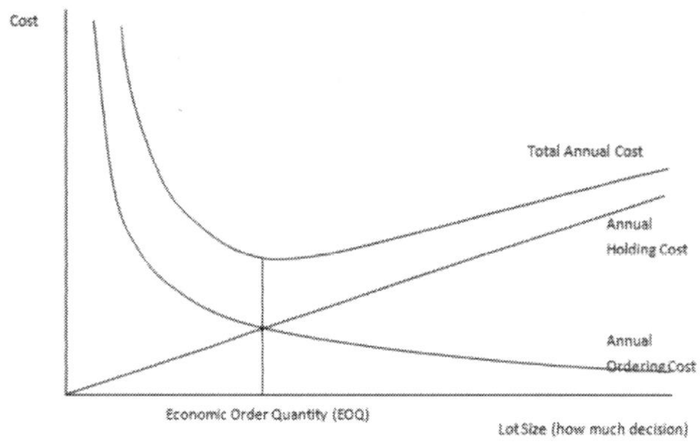

Fig. 1 Economic Order Quantity and Costs

2. Limitations of the Classic EOQ model

The assumptions made for the classic EOQ model result in many limitations for practical applications. Some of these limitations are –

i) Fixed unit cost

This assumption is reasonable when there is a long-term contract for supply of item and the per unit cost can be fixed. But in reality, we can have situations where there are price discounts based on quantity, items subject to inflation etc. which are not addressed. Delayed payment terms provided by suppliers can also help in reducing the total cost for the firm.

ii) No restrictions on order quantity

This assumption does not account for situations where perishable items are being handled. The inventory needs to closely match the demand to avoid wastage. Also, there could be warehouse / storage space restrictions that could put a cap on the order size.

iii) Stockouts are not allowed

This assumes that all the demand can be met always. But there can be cases where it is beneficial to allow backordering if the backordering costs are substantially low. Also, this situation happens when there is variable demand with time or any unexpected delays in supply etc.

iv) All items are perfect, no wastage

This assumption does not address the case when there are imperfect items produced due to various production constraints, perishable items remaining in inventory beyond their shelf-life etc.

3. Extensions to the Classic EOQ model

In the earlier section, we had looked at some of the limitations that were present in the classic EOQ model because of the assumptions. Many researchers have worked on extensions to the original model to address these limitations as needed. We will now look at how a few of these model extensions were done.

3.1. Fixed Unit cost limitation

Suppliers at times provide a price discount or delayed payments to the retailer so that it gives them an incentive to buy more. Biswajit Sarkar (2012) [3] presented a model that provides for optimal replenishment quantity under delayed payment terms along with stock dependent demand and in the presence of imperfect items. The author was able to establish a model that helped in optimizing cost based on the terms for delayed payment segmented into 3 distinct intervals. This kind of model is specifically applicable to goods like sugar, clothes and spices etc. that have stock dependent demand. Ata Allah Taleizadeh et. al. (2013) [4] developed an extension model that would address the case of a known price increase in the presence of partial backordering. In case of a permanent increase or temporary reduction in price of an item, the buyer can place a special order that is bigger in size than normal to take advantage. This model covers this aspect and helps in calculating the optimal order quantity for such a situation when the price change happens under inventory available or under shortage conditions. Sheng-Chin Chen et. al. (2013) [5] developed a simple arithmetic-geometric method instead of the differential calculus model to be applied to a retailer with permissible delayed payments from the supplier. They noted that the fraction of permissible delay payment was inversely proportional to the optimal reorder time, and the delay payment threshold was directly proportional to the reorder time.

3.2. No restrictions on order quantity

Even though the classic EOQ model placed no constraints on the order quantity, there are restrictions in terms of how much storage space a firm can afford. Many studies have been done in this respect to address these limitations. T.K. Roy et. al. (1997) [6] developed a fuzzy model that introduced fuzziness in both the storage space available as well as the unit price which was not constant, but varies inversely with the current demand. With their model, they were able to show better results than the usual crisp model.

In case of perishable items such as food or certain electronic items or chemicals, they usually have a shelf life after which they cannot be used for the original intended purpose. The problem could be due to various reasons such as decay, evaporation, loss of quality etc. Richard P. Covert et. al. (1973) [7] developed an extension that modelled the deterioration of the item over time using a Weibull distribution with assumptions of constant demand, no delay in delivery and no shortage. Their results were converging to other models that accounted for deterioration and also with the original model when the deterioration was negligible. Gede Agus Widyadana et. al. (2011) [8] worked on a model for the deteriorating items with and without backorder using a simplified approach. They used a modified cost-difference comparison for the same and the results were comparable to other models, and hence provides for a mechanism that is easier to solve. Cinzia Muriana (2016) [9] presented a mathematical model that had stochastic demand for perishable items with fixed shelf life. The varying demand was modelled using a normal distribution, and the model was used to determine the probability of having a perished item in inventory and based on this determine the optimized cost conditions. This helped in providing a better retailer strategy based on the parameters.

3.3. Stockouts are not allowed

Stockouts happen when there is customer demand, but the firm does not have enough items in inventory to meet the demand. This can lead to two possibilities – lost sales for those customers who do not want to wait, and backorders where the customers are willing to wait for stock replenishment, but potentially adding cost to the firm. Even though stockouts are costly and not a desirable condition, it is very much a reality and firms need to address them in their inventory management through partial or full backorders or accounting for lost sales.

David W. Pentico et. al. (2007) [10] modelled a deterministic EOQ that accounted for partial backordering. The model for a partial backorder is more difficult to approach than a full backorder or lost sales case and, in this respect, their model provided an easy approach to get the EOQ under such conditions. Their results also matched with those of full backordering and the basic EOQ model under similar conditions. N. Kazemi et. al. (2009) [11] developed a fuzzy model with backorders using triangular and trapezoidal fuzzy numbers for the input parameters and decision variables. This approach provided a better method than using a deterministic one. Kuhn-Tucker conditions were used to determine the optimal policy in this method. They established a directly correlated relationship between the decision variables and the differences in cost between the crisp and their fuzzy model.

Sujit Kumar De et. al. (2016) [12] approached the problem of backordering by considering promotional efforts such as price discounts, free gifts, delay in payments or better services etc. under stockout conditions. They developed a model with pentagonal strong / weak fuzzy variables applied to the classical backordered EOQ model with shortage time dependent demand rate and in presence of promotional efforts.

3.4. All items are perfect, no wastage

Even with the constantly improving quality control measures in firms, it is unrealistic to assume that all produced items would be perfect and no wastage or rework needs to be done. So, it is essential to incorporate the quality aspect into the classic EOQ model to mimic situations that are closer to real world.

M.K. Salameh et. al. (2000) [13] provided a framework for the EOQ model that would account for imperfect items. Their model used a 100% screening to identify the items that were imperfect and these were separated and subsequently sold at a discounted price for alternate use. So, screening cost became a significant cost in their model. But full inspection is very costly and may not be a reasonable approach. Building up on this, Jafar Rezaei (2016) [14] worked on a EOQ model that used sampling inspections for identification of imperfect items. Based on pre-determined thresholds, there were 3 options based on sampling – no inspection of the lot if the number of imperfect items is lesser than a lower threshold and these can be addressed using replacements eventually, full inspection if the sample value is between the lower and upper thresholds and finally reject the lot if the sample value is higher than the upper threshold. With this approach, the firm gets a better optimized cost and EOQ as it accounts for imperfect items.

Amir Hosein Nobil et. al. (2020) [15] extended the imperfect item study to determine the reorder point, which was missing from earlier such studies. Their research helped in incorporating certain rates of imperfect items, system costs and lead time to get maximum profit and efficiency.

Reference	Assumptions carried from Classic EOQ model	Assumptions varying from Classic EOQ model	Input / Decision variable pattern
Harris FW, 1913 (1), RH Wilson, 1934 (2) **Classic EOQ model**	Zero Lead time Infinite Replenishment Shortages not permitted No defective items Constant Unit purchase cost Payment at time of order delivery Fixed and finite demand		Deterministic
Biswajit Sarkar, 2012 (3)	Zero Lead time Shortages not permitted	Delayed payments Stock dependent demand Presence of imperfect items Finite replenishment rate	Deterministic
Ata Allah Taleizadeh, 2013 (4)	Payment at time of order delivery Constant Unit purchase cost Infinite replenishment	Known Price Increase Shortages allowed – Partial Backordering Zero cost of inspection	Deterministic
Sheng-Chih Chen, 2013 (5)	Constant demand rate Zero lead time Shortages not permitted	Permissible delay in payment	Deterministic
TK Roy, 1997 (6)	Zero lead time Shortages not permitted Infinite replenishment rate	Demand dependent Unit coast Limited storage capacity	Fuzzy
Richard P Covert, 1973 (7)	Constant demand rate Zero lead time Shortages not permitted	Imperfect items – in Weibull distribution	Deterministic
Gede Augus Widyadana, 2011 (8)	Zero lead time Fixed replenishment rate Constant demand rate No repair or replacement	Deteriorating item Shortages allowed – Planned backorder level	Deterministic
Cinzia Muriana, 2016 (9)	Fixed Lead time	Perishable products Fixed Shelf life Shortages allowed – demand lost Variable demand rate	Stochastic demand
David W Pentico, 2007 (10)		Shortages allowed – Partial and Full backorders, Lost Sales	Deterministic
N Kazemi, 2010 (11)		Shortages allowed – Backorders	Fuzzy (Input + Decision variables)
Sujit Kumar, 2016 (12)	Zero lead time Finite order size	Shortages allowed – Backlogging Price – Promotional Effort Selling price dependent on demand rate	Fuzzy
MK Salameh, 2000 (13)	Shortages not permitted	Imperfect quality items Reworked and sold as single lot	Deterministic
Jafer Rezaei, 2016 (14)	Shortages not permitted Zero lead time	Imperfect quality items Sample Inspection thresholds	Deterministic
Amir Hossein Nobil, 2020 (15)	Shortages not permitted	Imperfect quality items	Deterministic

Table 1 – Assumptions of Classic and Extension Reference EOQ models

4. Future work

As per details given in Table 1 and discussions earlier in this paper, many studies have been done to address the limitations of the classical EOQ model and provide extensions to them. But even with these extensions, most of them provide for deterministic input parameters and decision variables. There have been a few studies that have incorporated fuzzy patterns to these variables for more realistic modelling given their indeterministic nature. With all the various extensions, it provides for lot of interesting opportunities to work on fuzzy models for these to get better insights. The combination of various assumptions would also provide for research activities for specific applications.

References

[1] Harris FW 1913 How many parts to make at once. Factory, The Magazine of Management **10(2):**135-136, 152

[2] R. H. Wilson, 1934 A Scientific Routine for Stock Control, Harvard Business Review, **Vol. 13**, No. 1, pp. 116-128

[3] Biswajit Sarkar, 2012 An EOQ model with delay in payments and stock dependent demand in the presence of imperfect production, Applied Mathematics and Computation, **Vol. 218**, Issue 17, pp. 8295-8308

[4] Ata Allah Taleizadeh, David W. Pentico 2013 An economic order quantity model with a known price increase and partial backordering, European Journal of Operations Research, **Vol. 228**, Issue 3, pp. 516-525

[5] Sheng-Chih Chen, Leopoldo Eduardo Cardenas-Barron, Jinn-TsairTeng 2013 Retailer's economic order quantity when the supplier offers conditionally permissible delay in payments link to order quantity, International Journal of Production Economics, **Vol. 155**, pp. 284-291

[6] T K Roy, M Maiti 1997 A fuzzy EOQ model with demand-dependent unit cost under limited storage capacity, European Journal of Operational Research, **Vol. 99**, pp. 425-432

[7] Richard P Covert, George C Phillip 1973 An EOQ Model for Items with Weibull Distribution AIIE Transactions, **Vol. 5**, Issue 4, pp. 323-326

[8] Gede Augus Widyadana, Leopoldo Eduardo Cardenas-Barron, Hui Ming Wee 2011 Economic order quantity model for deteriorating items with planned backorder level Mathematical and Computer Modelling, **Vol. 54**, Issue 5-6, pp. 1569-1575

[9] Cinzia Muriana 2016 An EOQ model for perishable products with fixed shelf life under stochastic demand conditions, European Journal of Operational Research **Vol. 255**, Issue 2, pp. 388-396

[10] David W Pentico, Matthew J Drake 2007 The deterministic EOQ with partial backordering: A new approach, European Journal of Operational Research **Vol. 194**, Issue 1, pp. 102-113

[11] N Kazemi, E Ehsani, M Y Jaber 2010 An inventory model with backorders with fuzzy parameters and decision variables, International Journal of Approximate Reasoning, **Vol. 51**, Issue 8, pp. 964-972

[12] Sujit Kumar De, Shib Sankar Sana 2016 An EOQ model with backlogging International Journal of Management Science and Engineering Management, **Vol. 11**, Issue 3, pp. 143-154

[13] M K Salameh, M Y Jaber 2000 Economic production quantity model for items with imperfect quality, International Journal of Production Economics, **Vol. 64**, pp. 59-64

[14] Jafar Rezaei 2016 Economic order quantity and sampling inspection plans for imperfect items, Computers and Industrial Engineering, **Vol. 96**, pp. 1-7

[15] Amir Hossein Nobil, Amir Hosein Afshar Sedigh, Leopoldo Eduardo Cardenas Barron 2020 Reorder point for the EOQ inventory model with imperfect quality items, Ain Shams Engineering Journal, **Vol. 11**, pp. 1339-1343

Journal of Physics: Conference Series **2332** (2022) 012020 doi:10.1088/1742-6596/2332/1/012020

Computational Simulation of the Effect of Increase in the Estrogen level on the Normal- Tumour- Immune- Unhealthy Diet Model for breast cancer

Shankar Narayan S[1], Nidhi S Vaishnaw[1], Shivakumar B N[1], Aastha Vaishnav[2]

[1]Department of Mathematics, CMR Institute of Technology, Bengaluru- 560037
[2]Rugta College of Dental Science and Research, Bhilai-490024

nidhi.v@cmrit.ac.in

Abstract. Previous research findings have shown that an unhealthy diet can trigger the progression of tumors leading to a high mortality rate. Also, it has been noted that the hormonal changes caused in the human body support the proliferation rate in several kinds of cancer. In the present research, the light is thrown on breast cancer. Estrogens are a class of hormones that play an essential function in women's proper sexual and reproductive advancement. Moreover, studies have revealed that a woman's risk of breast cancer is proportional to the Estrogen and testosterone produced by the. Long-term and/or high-level exposure to these hormones has been associated with an increased risk of breast cancer. A novel mathematical model is framed to analyze the effect of estrogen levels on cancer advancement in females integrated with the effect of the unhealthy diet model. The levels of Estrogen are fixed to explore the other cell densities. The research concluded that the level of estrogen produced has a remarkable effect on the immune system in the case of a patient with breast cancer consuming an unhealthy diet.

1. Introduction

Breast cancer is a malignancy that causes abnormal cell proliferation in the breast tissue. Some recognized factors, including as hormones, genetics, and lifestyle, have the ability to enhance someone's chance of developing this condition. Estrogen is a crucial hormone for breast tissue growth, but it also contributes to breast cancer by causing DNA damage by creating oxidative metabolites. The discovery that estrogen-induced mutagenesis is linked to the recruitment of non-cycling, G0, cells into the cell cycle and a faster rate of progression through the G1 phase has shifted attention to estrogenic modulation of molecules involved in G1S phase progression. These findings show that estrogens influence c-Myc and cyclin D1 production and function, as well as stimulate cyclin E-Cdk2 complexes, both of which are rate-limiting for progression from the G1 to S phase.

Non-steroidal antiestrogens with a triphenylethylene-based chemical structure show a good prospect for treating hormone-dependent breast cancer. Tamoxifen (trans-1 - (4—dimethylamino ethoxy phenyl)-1.2-diphenyl but-1-ene) is now frequently used to diagnose advanced breast cancer in women. Tamoxifen's therapeutic efficacy is linked to the presence of estrogen receptors in breast tumors [1]. Furthermore, these nonsteroidal drugs have been shown to inhibit the growth of human breast cancer cell lines that contain estrogen receptors and to cause the regression of hormone-dependent mammary tumors in experimental mice. Antiestrogens are thought to attach to the estrogen receptor in the target cells. However, the exact mechanism of anti-estrogen activity remains unknown.

Estrogens are also powerful mitogens in a variety of tissues, including the mammary gland, where they play a key role in the genesis and progression of breast cancer. Many mathematical models represent the dynamics of normal cycling, many of which are based on Schlosser and Selgrade's formulation [2,3]. These models predict mean hormonal balance on a day-by-day basis [4,5]. There have been not many mathematical models adapted to predict the contraceptive effects of Estrogen on other hand Glenda et al. [6] and Margolskee and Selgrade [7] did not include ovarian autocrine effects in their initial menstrual cycle models, therefore they can't predict the contraceptive response to exogenous progestin injection.

In our study of the model, which includes estrogen effects as a compartment, we explore optimal control theory to analyze and control a condition. This is a typically used strategy to establish a method that ensures a minimum of unhealthy diet by following a fiber-rich diet. High fiber diets have been demonstrated to support reducing estrogen levels and also by reducing extra body fat, limiting refined carbs and processed meals, and limiting alcohol use, among other things.

In section 2 of the manuscript, we intend to formulate a novel mathematical model and analyze it mathematically for its solution existence and optimization of certain terms in the next section. In section 4, we present the results obtained by solving the model using numerical techniques. A detailed description of Tumor-Immune-Estrogen interrelation has been explored in this section of the manuscript. Finally, we conclude the work by summarizing the key findings of our research in the form of the final section of the manuscript- Conclusions.

2. Mathematical Model Formulation

This part looks at the best ways to govern a mathematical model of tumor growth and its interaction with the immune system, as well as the impact of a healthy lifestyle. N-Normal cells, T-Tumor cells, I- Immune cells, and E- Estrogen levels are the four compartments in this model. In this work, we looked at a mathematical model.

$$\frac{dN}{dt} = rN(1-\beta_1 N) - \eta NI - \gamma NT - (1-k)\phi_1 NE$$

$$\frac{dT}{dt} = \alpha_1 T(1-\alpha_2 T) + \beta_2 NT - \alpha_3 TI + (1-k)\phi_1 NET$$

$$\frac{dI}{dt} = \sigma - \delta I + \frac{\rho NI}{m+N} + \frac{\rho_1 TI}{m_1+T} - \mu NI - \mu_1 TI - (1-k)\phi_2 IE$$

$$\frac{dE}{dt} = (1-k)\varepsilon - \beta_4 E$$

The block diagram of the model formulated is presented in figure 1.

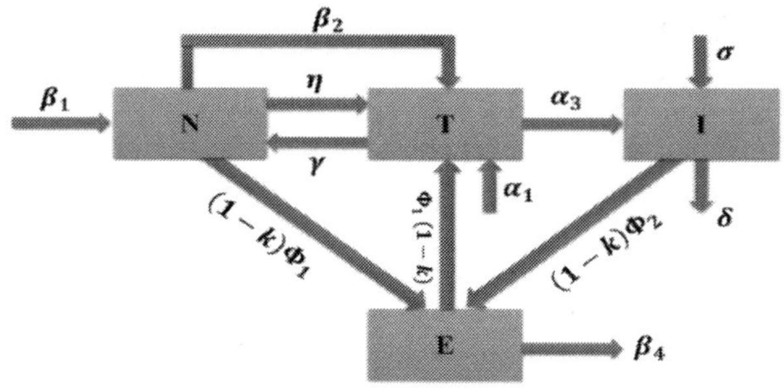

Figure 1. Block diagram of the mathematical model

3. Mathematical analysis of the model

The function $u \geq 0$ represents the immune cell rate. Positive values of u correspond with an influx of DCs supplied from outside the system. Hence, equation for (I) becomes

$$\frac{dI}{dt} = \sigma - \delta I + \frac{\rho NI}{m+N} + \frac{\rho_1 TI}{m_1+T} - \mu NI - \mu_1 TI - (1-k)\phi_2 IE + u$$

We now deploy optimal control theory by defining an objective function. The aim of the objective function is to minimize the tumor burden over time. Here we adopt the function u^2 as in similar work [8]. Hence we aim to minimize M[t, u],

$$M[t,u] = \int_0^t \left[T(t) + \frac{B}{2}u^2\right] \quad (1)$$

Where t is a specific time and B is a weighting factor that represents the relative importance of a healthy diet in daily routine which tense to the reduction of estrogen hormone. To proceed with the necessary condition of optimal control, we seek to rewrite and solve the governed system as follows

$$y' = k(t,y,u)$$

Where $x(t) = [N(t), T(t), I(t), E(t)]^T$ is the vector of the solution of the system equations, and k is the vector of the right-hand sides of the equations. Also, we have

$$M[t,y(t),u(t)] = \int_0^t [g(t,y,u)]\,dt,$$

Where g is the integrand of the equation (1).

We apply Pontryagin's Principle [9] and define the Hamiltonian function of this optima

$$H(N,T,I,E) = [g(t,y,u)] + \lambda k(t,y,u)$$

We have the necessary optimal condition

$$-\frac{\partial H}{\partial u} = 0 \quad at \quad u^* \Rightarrow g_u + \lambda k_u = 0$$

The partial derivatives as the adjoint equation are given by

$$\lambda' = -\frac{\partial H}{\partial x} = -(f + \lambda g), \qquad \lambda(t) = 0.$$

We now consider the optimal control problem

$Min \int_0^t g(t,y,u)$, subject to $y' = k(t,y,u)$ with initial conditions $x' = g(t,x,u), x(t_0) = x_0$.

Table 1. Parameter values and initial values

Parameter (unit)	Description	Value	
r (day^{-1})	Growth rate	**4.31×10^{-1}**	
β_1 (cell day^{-1})	Inhibition rate of *N*	**2.99×10^{-6}**	
η(cell^{-1}day^{-1})	Immune cells that inhibit or eradicate abnormal cells	**0.1379**	
γ (cell day^{-1})	The rate at which tumor cells inhibit or attack normal cells	**0.9314**	
k(day^{-1})	Effectiveness of anti-cancer drugs	**0 to 1**	
ϕ_1(day^{-1})	Tumor formation rate resulting	**4.12×10^{-2}**	
α_1 (cells)	Limited growth of tumor cells	**4.426 ×10^{-1}**	
α_2 (cells)	Confronting a decline in tumor cells	**0.4**	
α_3 (cells)	Rate of inhibition or the eradication of tumor cells by immune cells	**0.1469**	
β_2 (cell^{-1}day^{-1})	Rate at which abnormal cells	**1.1890**	
σ(day^{-1})	A constant source of the immune system response	**0.7**	[10,11]
δ(day^{-1})	The natural rate at which immune cells die	**0.57**	
ρ(day^{-1})	Source rate of immune cells	**0.2710**	
m (day^{-1})	Immune system's threshold rate	**0.8130**	
ρ_1(day^{-1})	Parameter affecting the response of the immune system	**07829**	
m$_1$ (day^{-1})	Parameter affecting the response of the immune system	**0.8620**	
μ(day^{-1})	Reduced immune cells rate	**0.8130**	
β_4 (cell day^{-1})	The natural death rate of estrogen	**0.97**	
μ_1 (day^{-1})	Logistic rate of normal cells	**0.3634**	
ϕ_2(day^{-1})	Immune suppression rate due to excess estrogen levels	**0.002**	
ε (day^{-1})	Source rate of estrogen	**1.5**	
	INITIAL VALUES		
T	Tumor cell population	**10^{6}**	
N	Normal cell population	**10^{5}**	
I	Immune cell population	**10^{4}**	
E	Estrogen levels	**10^{-4}**	

4. Results and Discussion

Failure of the immune system during cancer development is caused due to the unhealthy diet followed by the patient. Also, lifetime exposure to the estrogen produced by the own body is responsible for the promotion of breast cancer to the next stages. The mathematical model designed in the previous section is solved numerically through a FEM solver. The parameter values given in table 1 are fed into the solver and the system-generated graphs are mentioned in the present section to analyze the effect of an unhealthy diet and the estrogen level. The graph of various cell densities with time is explored to investigate the effect of an unhealthy diet and estrogen levels.

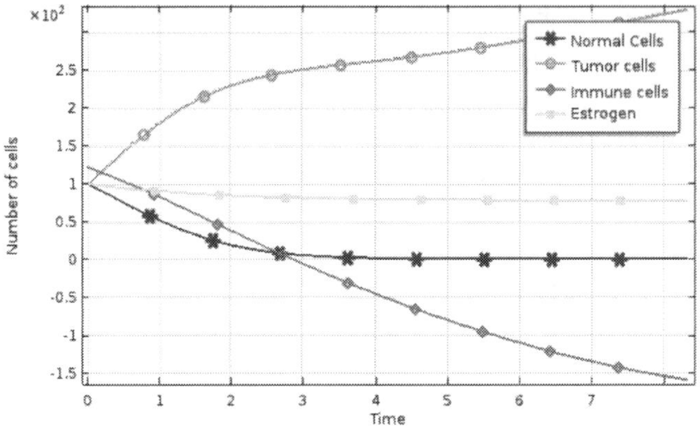

Figure 2. Graph of various cell densities under study

Figure 1 represents the overall cell densities of normal cells, tumor cells, immune cells, and estrogen levels. For the source rate of estrogen considered, with the initial level of estrogen as 10^{-4}, the plots suggest that the 1.5 source rate of estrogen is sufficiently large enough to promote the development of cancer in the breast. The growing graph of tumor cells suppresses the level of normal cells nearly exponentially promoting the growth of cancer in the patient, as seen in figure 2. For the initial conditions adopted, the growth of a tumor nulls the normal cell density within a short period of time.

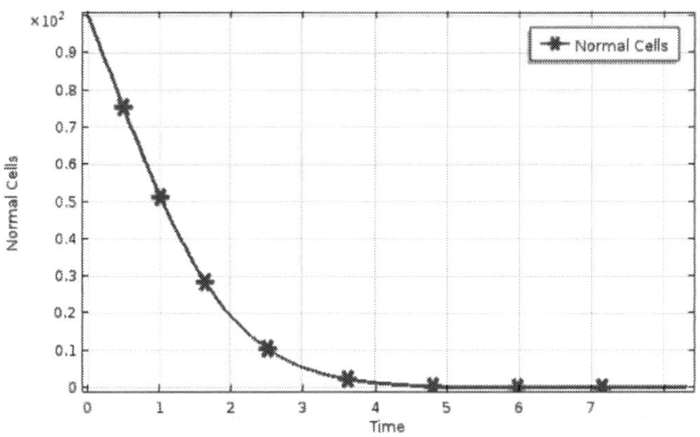

Figure 3. Normal cell density

Also, from figure 3, it is evident that the effect of estrogen supported by the unhealthy diet on the immune cell density is considerable and negative in nature. The loss of immune cells is majorly due to the unhealthy diet taken up by a person. This loss in the immune cells accelerates the rate of proliferation in the cancer cells leading to advancements in the symptoms. The fuel to this increase in the rate of proliferation is the hormone- estrogen. Few breast cells need the estrogen hormone for growth. The specific biological pathways by which estrogens promote breast tumor development are unknown [12]. There are several forms of breast cancer. 70% of them seem to be susceptible to the female sex hormone estrogen. These cancer cells have receptor sites that bind to estrogen, promoting their proliferation and spread. These are referred to as estrogen receptor-positive malignancies (or ER-positive cancers). Tumor cells are examined to check if they contain these receptors. If this is the case, hormone (or endocrine) therapy may be utilized to address the condition.

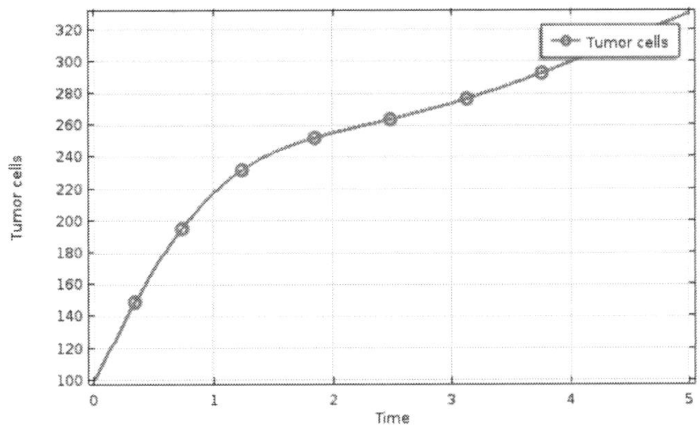

Figure 4. Tumor cell density

Figure 4 represents a progressive tumor cell density plot, where the progression is due to the terms representing the unhealthy diet and estrogen levels in the mathematical model formulated. The growth is accelerated due to the factors mentioned before. Figure 5 epitomizes the retarding immune cell population. The drop in the plot seen explains that estrogen levels can be elevated due to the unhealthy diet maintained by the patient. Whilst the impacts of this hormone have been characterized differently in each immune progenitor, it is known that cytokines connect all of them, maintaining a dynamic interaction in the tumor microenvironment [13]. Also, it has been proved that the cytokine level can be elevated by foods containing high sugar content [14].

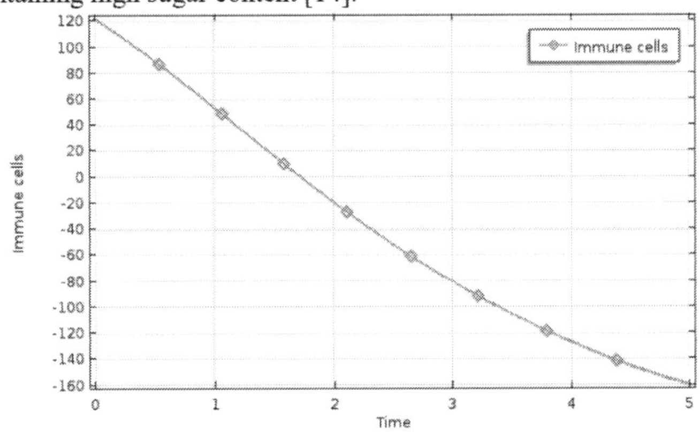

Figure 5. Immune cell density

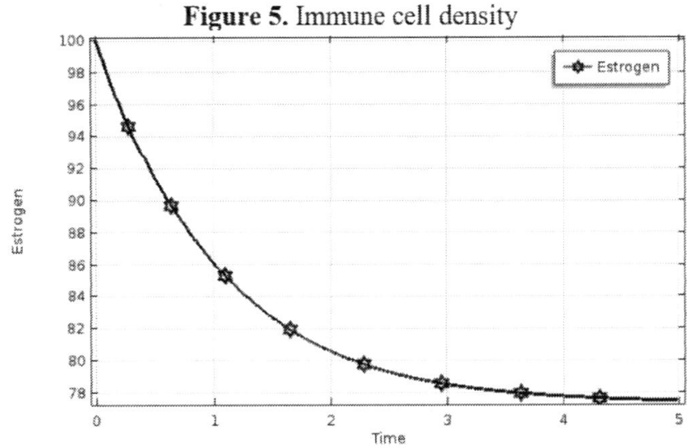

Figure 6. Estrogen level

Finally, figure 6 represents the estrogen decreasing pattern for the considered source rate of the hormone. It is seen that for the source rate of estrogen considered for the present study, the estrogen concentration tends to fade away with time causing remarkable changes in the immune cell, tumor cell, and normal cell densities. Considering the source rate of estrogen to be 1.5, the level of estrogen considered in the closed system represented by the mathematical model formulated causes: (i) accelerated normal cell proliferation resulting in the increase in the tumor cell density, and (ii) depleting the immune cell density.

Finally considering the cases for varied estrogen source rate, the following results are deduced based on the values of various cell densities presented in Table 2. From table 2, it is clear that the change in estrogen source rate affects the number of tumor cells. In fact, we see a rise in the tumor cell density as the value of $\boldsymbol{\varepsilon}$ rised.

Table 2. Cell densities with estrogen source rate

	$\varepsilon = 1.5\ per\ day$			$\varepsilon = 1.0\ per\ day$			$\varepsilon = 0.5\ per\ day$			$\varepsilon = 0\ per\ day$		
Time	**N**	**T**	**I**	**N**	**T**	**I**	**N**	**T**	**I**	**N**	**T**	**I**
0	100	100	122	100	100	122	100	100	122	100	100	122
1	52	218	54	52	219	54	52	218	54	52	218	54
2	20	255	-20	20	254	-20	19	255	-19	20	253	-20
3	6	274	-80	6	273	-80	6	273	-80	6	272	-79
4	2	300	-125	2	300	-125	2	300	-125	2	299	-126
5	0	330	-160	1	330	-160	1	330	-160	1	329	-159

5. Conclusions

An Unhealthy diet model coupled with the estrogen level compartment is designed to study the cancer development in breast cancer patients. The study concentrated on the effect of the values of source rate production of estrogen. The following conclusions are drawn at the end of the present study.

- The effect of an unhealthy diet coupled with the 1.5 estrogen level greatly affects the cancer development rate. The exponential drop in normal cells is seen due to the accelerating support given by the estrogen molecules to cancer development.
- An unhealthy diet integrated with the tumor-progressive effect of estrogen molecules demands hormonal therapy (also known as Endocrine therapy).

6. References

[1] Doisneau-Sixou S F, Sergio C M, Carroll J S, Hui R, Musgrove E A and Sutherland R L 2003 Estrogen and antiestrogen regulation of cell cycle progression in breast cancer cells *Endocr. Relat. Cancer* **10** 179–86

[2] De Pillis L G, Gu W and Radunskaya A E 2006 Mixed immunotherapy and chemotherapy of tumors: Modeling, applications and biological interpretations *J. Theor. Biol.* **238** 841–62

[3] Olden1 K and Newbold2 R R 1999 *Environmental Health Perspectives ?* vol 1081

[4] Garcia-Morales P, Saceda M, Kenney N, Kim N, Salomon D S, Gottardis M M, Solomon H B, Sholler P F, Jordan V C and Martini M B 1994 Effect of cadmium on estrogen receptor levels and estrogen-induced responses in human breast cancer cells *J. Biol. Chem.* **269** 16896–901

[5] Prall O W J, Rogan E M and Sutherland R L 1998 Estrogen regulation of cell cycle progression in breast cancer cells *J. Steroid Biochem. Mol. Biol.* **65** 169–74

[6] Gillies G E and McArthur S 2010 Estrogen Actions in the Brain and the Basis for Differential Action in Men and Women: A Case for Sex-Specific Medicines *Pharmacol. Rev.* **62** 155–98

[7] Margolskee A and Selgrade J F 2011 Dynamics and bifurcation of a model for hormonal control of the menstrual cycle with inhibin delay *Math. Biosci.* **234** 95–107

[8] Trisilowati T, McCue S and Mallet D 2012 Numerical solution of an optimal control model of dendritic cell treatment of a growing tumour *ANZIAM J.* **54** C664–80

[9] Bittner L 1963 L. S. Pontryagin, V. G. Boltyanskii, R. V. Gamkrelidze, E. F. Mishechenko, The Mathematical Theory of Optimal Processes. VIII + 360 S. New York/London 1962. John Wiley & Sons. Preis 90/– *ZAMM - J. Appl. Math. Mech. / Zeitschrift für Angew. Math. und Mech.* **43** 514–5

[10] Yousef A, Bozkurt F and Abdeljawad T 2020 Mathematical modeling of the immune-chemotherapeutic treatment of breast cancer under some control parameters *Adv. Differ. Equations* **2020**

[11] Alharbi S A and Rambely A S 2020 A new ODE-based model for tumor cells and immune system competition *Mathematics* **8** 1–14

[12] Yue W, Wang J P, Li Y, Fan P, Liu G, Zhang N, Conaway M, Wang H, Korach K S, Bocchinfuso W and Santen R 2010 Effects of estrogen on breast cancer development: role of estrogen receptor independent mechanisms *Int. J. Cancer* **127** 1748

[13] Segovia-Mendoza M and Morales-Montor J 2019 Immune tumor microenvironment in breast cancer and the participation of estrogens and its receptors into cancer physiopathology *Front. Immunol.* **10** 1–16

[14] Quigley D A, Tahiri A, Lüders T, Riis M H, Balmain A, Børresen-Dale A L, Bukholm I and Kristensen V 2017 Age, estrogen, and immune response in breast adenocarcinoma and adjacent normal tissue *Oncoimmunology* **6**

AUTHOR INDEX